AF254616

Family With Six Treasures

家有六仟金

Li's Family has Six Beautiful and Charming Daughters

李家有六位如花似玉、千嬌百媚的女兒

Gwen Li

余國英　著

美商EHGBooks微出版公司
www.EHGBooks.com

EHG Books 公司出版
Amazon.com 總經銷
2021 年版權美國登記
未經授權不許翻印全文或部分
及翻譯為其他語言或文字
2021 年 EHGBooks 第一版

ISBN-13：978-1-64784-072-3

Thank You

Edited by Wendy Chou (2019.10.30)

"How about I translate my Chinese articles into English so you can enjoy   them with ease?"   I   suggested to my English reading relatives and friends generosity. And also figure that it will be a very good chance for me to practice my own English as well.   Alas, It turned out that was so much easier said than done! After translating some of my published Chinese articles into English,   I certainly have to admit   that   the English really is not my first language! It does need a lot of editing to make the reader really enjoy the articles!

Whom should I ask to do this unprofitable job? Since I don't have any money to pay for their hard work?

For fear of too much works may take too much of their time, in order to reduce the burden of each person's work, I have asked more than one to share   the editing work, they are my dearest friends or relatives:

1, **Adela Karliner,** In-law, college English professor, Retired lawyer.

2, **Sandy Chin**, Friend, Special Educationer.

3, **Descartes Li**, son, professor of psychiatrists
4, **Wendy Chu**, niece, environmental worker.

5, **Ruth Yen**, niece, Librarian.

6, **Pearl Karliner Li**, granddaughter,

I would like to express my sincere gratitude to all six of them for the time, effort and knowledge they have taken, in order to polish my work!

I have always thought that all   the people in the world are   all my friends, and those who are willing to read my hard work which were written with my   sweat and tears in Chinese, or willing to read the my English article which had been edited by my friends and relatives are all my dearest "best friend"!

Here, I sincerely invite all my best friends to my blog:

http://blog.udn.com/gwenli2013/article
http://www.overseaswindow.com/home/writer/1129
http://www.amazon.com/author/gwen.li
http://wxs.hi2net.com 余國英

Again, Thank you all very much!

感謝

　　「那，這樣好了！我把發表过的中文章翻譯成英文再給不讀中文的讀者们看，不就行了嗎？」為了讓不能讀中文的讀者讀我的文章,我自告奮勇，非常慷慨地回答，心想這樣也給了我鍛練自己英文的機会。

　　不過，英文到底是我的第二語言，伊妹兒給自己人看一看，笑一笑就罷了，要轉寄給大家看，只怕我的英文文字不夠完美，有辱大家的視聽，就央求那些英文修養比較好的親戚朋友們，替我把英文文章編輯一下再發出去，變成不定期的轉寄，沒有想到數年下來，竟然可以積成一本書，真是太高興了！

　　因為怕太打擾工作繁忙的親戚朋友們，為了要把每人工作擔子減輕一些，所以請求了不止一位來改正編輯，以期減輕各人的工作量，這些人按次序排列，包括：

1，**柯艾達**，親家，律師及英文教授。

2，**秦珊蒂**，好友，特殊教育者。

3，**李德康**，兒子，心理系副主任醫師及教授。

4，**周文婷**，姨姪女，環境工作者。

5，**顏如詩**，外甥女，圖書館理員。

6，**李維珍**，孫女兒，大學畢業。

在此，我誠懇地向上述各位致以深深的謝意，因為佔用了他們寶貴的時間，心力以及他們的知識，而且沒有薪酬，才能有本書的完成。

　　嘻嘻，我一直認為世人皆吾友，而願意讀我用汗和淚寫出來的小說的中文讀者，或者願意讀過高手編緝過我的英文文章，更是我最最親愛的「好朋友」喲！

誠懇遊請我的好朋友們到我的部落格來交換意見：

http://blog.udn.com/英姐的部落格

海外文軒/余國英

文心社/余國英

Acknowledgments

Adela Karliner: In-Law, College English Professor, Retired Lawyer.

Dr. Descartes Li is *Vice Chair for Education, Department of Psychiatry, Weill Institute for Neurosciences*, San Francisco. He gives numerous talks and presentations in both academic and community settings, and has a wide range of interests that include bipolar disorder, medical education, electroconvulsive therapy, cultural psychiatry, and suicide prevention. He was born in New York, and grew up on Long Island. A graduate of Columbia College of Columbia University, he now lives and works in San Francisco, California. He is the son of the well-known Chinese American author, Gwen Li.

Dr. Wendy Chou is a science communicator with a strong background in climate change. She is currently Communications and Outreach Manager for the Bay Area environmental nonprofit Acterra. She has experience writing and consulting for varied audiences including higher education, scientific journals, public agencies, community-based organizations, and the private sector. Wendy received her Ph.D. in Environmental Science, Policy, and Management from the University of California – Berkeley and holds a B.A. from Harvard.

Ruth Yen Wong is a middle school library media specialist who is an avid reader, loves tennis, and gardening. She grew up in Columbus, Ohio, but now lives in southeast Michigan. She has a wonderful husband and two sons. A graduate of Ohio State University with a BS in computer information science, she was a computer programmer out of school and later held positions as Cub Scout den mother, ESL tutor, and K-8 substitute teacher before finding her way to the library. Ruth is also the niece of the talented author, Gwen Li.

Professor Leah Karliner

Pearl Karliner-Li is a recent graduate of Barnard College in New York City. She currently lives in San Francisco close to where she grew up and is happy to be so close to her Nai Nai in Oakland and to be able to assist her with her writing. She also loves rock climbing, reading, doing yoga, cooking (and eating), and doing ceramics.

Table of Contents 目錄

我一直認為，世人皆吾友，而願意閱讀我用汗和淚寫出來的作品的讀者們，若能更進一步地踴躍發表寶貴的意見，能使我更上層樓，更是我最最親愛的好朋友喲！

誠請我的好朋友們到我的部落格來交換意見：

wxs.hi2net.com/home/blog.asp?id=184

blog.udn.com/gwenli2013

www.overseaswindow.com/user/1129

希望各位讀者喜歡，並且敬請上網搶購！

余國英

✉ 聯絡作者

粉絲專頁 https://zh-tw.facebook.com/public/余國英

購書專頁 http://www.amazon.com/author/gwen.li

Foreword

Edited by Wendy Chou(2019.10.30)

Having Lived in the United States for more than half a century, I have been using Chinese traditional and simplified characters to write the stories of Chinese compatriots in the U.S., Quite simply, because I was born and lived my formative years in China, so Chinese is my   first language.

My son and his wife told me that there have been times when they were eating at a Restaurant, and a manager or waiter came over to say, "Yesterday, in the newspaper we read your mother's article in which she mentioned you!"

"That's great!   Can I borrow your paper?" My son would answer.

When the newspaper was brought over, they found it was in Chinese; "Mom, you know, although I've studied Chinese for many years, but I can't read Chinese newspapers and magazines as well as I read English!" My son complained to me.

The same thing happened to the next generation of my American–born relatives, including my nephews, nieces, and their families and friends as well as my grandchildren.

Deep down in my heart, I do perfectly understand his feelings! Like many naturalized-Chinese Americans, even though I have lived in the U.S. for more years than I lived in China, however, the Chinese language still comes to me lots easier than English does.

To reach all these potential readers, I decided to translate my works into English. And I figured that it would also be a very good chance for me to practice my English. Alas, it turned out to be easier said than done! When I read the translated English version of my article, I suddenly realized that English certainly is my second language! It does need a lot of editing to help the reader really enjoy the article!

For fear that of too many works may take too much of their time, (since after all, I did not have any money to pay any of them), and in order to reduce each person's share of burden, I have asked these dear friends or relatives to share the editing work:

My friend and in-law, **Adela Karliner**; my friend, **Sandra Chin;** my son, **Descartes Li;** my niece, **Wendy Chu**; **Ruth Yen;** and my granddaughter, **Pearl Karliner Li**.

I didn't imagine that in a few years, it would all be accumulated into a book, but it certainly has made me very happy!

I would like to express my sincere gratitude to all of you for the time, effort and knowledge that you have contributed to help this book reach completion.

The original articles of this book were published in Chinese, later translated into English by the author, -and the English text were further edited and polished by Sandra, Descartes and Wendy. -Therefore, the original Chinese versions which have been included at the end of the book for reference, should not be thought as word-for-word "English-Chinese parallel texts." it should be more interesting to read the way it is.

I have always thought that all the world's people are all my friends, those who are willing to read in Chinese, the work which I created through sweat and tears, or are willing to read in English those words which have been edited by my expert friends and relatives, are all my dearest "best friends"!

I sincerely invite my all my best friends to my blog to exchange views:

http://blog.udn.com/gwenli2013/article

http://www.overseaswindow.com/home/writer/1129

http://www.amazon.com/author/gwen.li

http://wxs.hi2net.com 余國英

www.overseaswindow.com › home ›余國英

自序

　　在美國居住已經超過半世紀，一直用中文的繁體字及簡体字來書寫中國同胞在美國異鄉的故事，因為中文是我這在中國出生的人的第一文字啊！

　　據兒子和媳婦說，有幾次，他們在餐館吃飯，餐館的經理或服務員居然跑過來對他們說：「昨天的某某報紙或某某雜誌看見你們家奶奶的文章，裡面還提到了你們呢！」

　　「太好了！可以借來看看嗎？」我兒子告訴我他是如此回答他們的。

　　等他們拿了報紙來一看，原來是中文的。「媽，妳是知道的喲，我雖然學了幾年中文，但是要我讀中文報紙和雜誌還不太靈光的呀！」兒子向我訴苦。這個苦衷我實在太了解了！因為我在美國住的日子比在中國多了去了，可是，我讀英文還是得一句一推敲，不像讀中文可以一目十行，何況要把自己的意思用英文表達出來呢？

　　孫子及孫女兒也對我說過：「我的朋友的中國媽媽或者爸爸也跟我說看見過奶奶在報紙雜誌上的文章，可惜哦，我們和我們的朋友或同學，都看不懂中文耶！」其他在美國出生的姪兒、姪女、外甥、外孫女和他們家的人都紛紛表示同意呢！

　　「那，這樣好了！我把他們翻譯成英文再給你們看，不就行了嗎？」我立刻自告奮勇，非常慷慨地回答，心想這樣也給了我鍛練自己英文的機会呢。

　　不過，英文到底是我的第二語言，伊妹兒給自己人看一看，笑一笑就罷了，要轉寄給大家看，只怕我的英文文字不夠完美，有辱大家的視聽，就央求那些英文修養比較好的親戚朋友們，替我把英文文章編輯一下再發出去，變成不定期的轉寄，沒有想到數年下來，竟然可以積成一本書，真是太高興了！

　　因為怕太打擾工作繁忙的親戚朋友們，為了要把每人工作擔子減輕一些，所以請求了好幾位來改正編輯，以期減輕各人的工作，這些人按年齡排列，包括：

　　1, 柯艾達，親家，英語教授。

　　2, 秦珊蒂，好友，特殊教育者。

　　3, 李德康，兒子，心理醫師教授。

　　4, 周文婷，姨姪女，環境工作者。

　　5, 顏如詩，外甥女，圖書管理員。

6, 李維珍，孫女，辦公室工作者。

在此，我誠懇地向上述各位致以深深的謝意，因為佔用了他們寶貴的時間，心力以及他們的知識，才能有本書的完成。

本書的文章在報紙雜誌上發表的是用中文寫的，又經作者翻譯成英文，再加上英文文字經過高手的潤色，一定更加精彩。"中英文對照"是把中文的原文放在英文的後面當作參考，想來也是別開生面很有意思的。

嘻嘻，我一直認為世人皆吾友，而願意讀我用汗和淚寫出來的小說的中文讀者，或者願意讀過高手編缉過我的英文文章，更是我最最親愛的「好朋友」喲！

誠懇遊請我的好朋友們到我的部落格來交換意見：

http://blog.udn.com/英姐的部落格

文心社/余國英

海外文軒/余國英

My God, You Have Even Changed Your Name!

Edited by Sandra Chin

Today, as I read my newly published book in the Evergreen Club, Grandpa Zhang tottered falteringly by with his cane.

"Hi, Grandpa Zhang, hello," I greeted him politely as I put down my book.

"Nancy, honey, long time no see.   You have changed a lot!" he said in English as he narrowed his eyes in surprise.

"Grandpa Zhang, we just had breakfast together!" I protested.   "And, by the way, my name is Gwen!"

"Heck, you even changed your name!" He was even more shocked.

"WanJun, you should give your father a notebook to jot down everything so he won't forget," I suggested to his daughter who followed closely behind her father.

"I have done just that, but I forgot where I put the notebook," Wanjun   answered.   She is also a retired senior.

"Did you know there is a drug called ginkgo biloba that can help your memory?" I suggested.

"But His Holiness forgets to take it!"

"Please don't think we seniors have bad memories! See, I do remember the 7 essentials.   When you open your door daily, there are 7 things you should worry about. Look, the seven things are fuel, rice, oil ….!" Grandpa Zhang said loudly, but then stopped short.   He must have forgotten the other 4 essentials.

"Dad, when you get up in the morning, what are the seven pills you should take?" his daughter asked.

"Well, I do remember! Take the vitamins,   swallow cod liver oil, and ….!" Ah, he could not remember!

Maybe we better leave things as they are, I said to myself. At least, he called me   "Nancy honey,"   in English, rather than   "Doncy NoSalt 東施、無盐"   in Chinese, the names of   two hideous ancient Chinese crones!

老天，連名字都改了

今天，我在長青俱樂部看我新出版的書，拄了柺杖的張老爹顫巍巍地走過來。

「張老爹，您好。」我向他打招呼。

「南施蜜糖，好久不見，妳改變了很多！」他覷著眼睛吃驚地用英語說。

「張老爹，我們剛才才在一起吃早飯，還有，我的名字叫國英。」我抗議道。

「哎呀，連名字都改了！！」他更吃驚了。

「婉君，替妳家老爹準備一個本子，凡做了任何事都隨手記下來，不就忘不了了嗎？」我向跟在他身後的女兒建議。

「是準備了，可是忘了放在那裡啦。」張婉君回答，她也已經退休了。

「對了，現在有一種藥，叫做 ginkobiloba，吃了可以幫助記憶呢。」我又出了一個好主意。

「可是他老人家忘了吃喲。」

「千萬別說老人家上了年紀記性就不好，看，開門七件事，柴米油 ⋯⋯」老爹大聲地說。

「對了，清晨起床，起床七件事是些什麼呢？」做女兒的問。

「哼，吞了維他命，嚥了魚肝油，還有，還有，⋯⋯ 想不起來了！

我想，還是不要追究罷，至少，他用英文叫我「南施蜜糖」，沒有用中文叫我「東施無鹽」，我的運氣已經算挺不錯的啦。

The Warmth in a Cold Day

Edited by Sandy Chin

It is December. The outside is cold and brisk, so this morning I leave my house a little later than usual for my daily walk. After walking for a while, my body gradually warms up. The bright sunshine in the cold weather makes the houses and trees on the street look particularly bright and fresh, and my heart feels warm and happy.

In the past we lived in the Florida countryside. During this time of the year, every night, every family had colorful lights lit inside their homes, on the eaves, and in the courtyards. Each whole place would be decorated with bright and colorful Christmas decorations and the lights would stay on until the sun came out the next morning.

Now, I have moved to California, and am living in Oakland's Chinatown, and I see that people are busy opening their stores to welcome the day's business. There is no sign of Christmas celebration. Ah, yes, they are waiting for the Chinese New Year. Then the lights will bloom and the big celebration will begin.

While I was guessing, I saw a chunky twenty-year-old white youth sitting alone on the stone chair outside the building, grinning at all the pedestrians coming and going through the corridor. His clothes were clean, he was quiet, and seemed well-mannered and pleasant, and he carried a packet of beautifully packaged candy.

"Merry Christmas! Hi, lady," he called to a passing kind-faced and nice-looking old lady. "Do you like to share a bag of candy with me?" He called her lady which meant they did not know each other.

"Well, young man, thank you, and I wish you a Merry Christmas too!" the old white lady answered happily. Here in Chinatown, white people can be counted as minorities.

I saw the white young man open the splendid, shiny candy package in front of the white old woman. Then they each one popped a piece of candy into their grinning mouths.

Before leaving, the old lady not only took out a red apple from her shopping bag to give to the young man, but she also gave him a huge, warm hug.

Seeing this touching scene take place under the bright winter sunshine made me, a Chinese passerby's heart feel particularly warm.

寒天裡的溫暖

　　已經十二月了，外面愈來愈冷，所以我今天清晨散步，也特地晚一些出門，走了不久，身子漸暖，明媚的太陽光照在冷冽的天氣裡，使得街道、房屋以及街邊的樹木，都顯得格外燦爛清新，心裡也覺得暖洋洋，喜孜孜。

　　以前我們住在鄉下，每到夜晚，各家各戶獨立家屋內、簷下以及庭院內外，所有五顏六色的綵燈，都一齊大放光明，有的一直到第二天太陽出來，才把聖誕裝飾用的燈光熄滅。

　　目前我已搬入奧克蘭的中國城內，此時看見各商家正在忙著打開店門做生意，并沒有什麼慶祝聖誕節的跡象，啊，對了，他們是在等待春節，中國農曆新年才張燈結彩，大肆慶祝罷。

　　心裡正在猜度，只見一個廿餘歲的白人青年，大概因為身體有點肥胖罷，獨自坐在大廈外面的石椅上，笑嘻嘻地望者走廊裡來來往往的行人，因為他衣著乾淨，行為也很斯文，所以并不引人反感，他手中拎著一小包包裝精美的糖果，對著路過的一位面目和善，舉止慈祥的白人老太太說：「聖誕快樂！這位女士，妳願意與我分享這包糖果嗎?」他叫她女士，可見他們并不相識。

　　「好呀!年青人，謝謝你，也祝你聖誕快樂!」白人老太太也很高興地回答。在我們這裡，白人可以算作少數民族。

　　只見那位白人青年，當著那位位白人老太太的面，把花花綠綠金光閃亮的包裝盒子撕開，二人一人一粒糖果，各自送入笑嘻嘻的口中。

　　臨走時，老太太不但由購物袋中取出一個紅蘋果送給這位青年，還送給他了一個溫暖的擁抱。

　　這冬日陽光下明亮燦爛的一幕，看在我這中華路人的眼裡，心裡覺得特別溫暖。

The most beautiful barber in the world

Edited by Adela Karliner (08/26/2020)

When she was four months old, her father, my son, took this picture.   They had lined up all the four-month-old babies on a sofa since none of them were capable of sitting up straight, so they all had to lean on something. The second one from the left, who is wearing her aunt's hand-knitted pink sweater, was my lovely granddaughter.

There is a Chinese saying that when a girl is growing up, she will go through eighteen changes. This statement is especially true when it applies to my granddaughter. Let's look at her now. She has beautiful, creamy, clean skin. And her thick shining auburn hair is just like her maternal grandmother's. She is smart enough to edit my English articles and is also a great painter. The wall of her home is full of her paintings. Once I went to her school auditorium to watch a play. A giant ceiling to floor fresco occupied a whole side of the auditorium wall. My daughter-in-law smiled and said to me, "Nai Nai, the painter's name is Pearl Karliner Li. She is your lovely granddaughter!"

During the COVID-19 pandemic prevention period, one must comply with the government's stay-at-home order. So to avoid infection, it is best not to go out. My son, who is the new vice-chairman of his department, only goes to the hospital once a week, and the rest of the time is on call at home. After her college graduation, my granddaughter found a computer trainee's job with The University of California, San Francisco School of Medicine. It will take three months before she could turn into a formal employee. She is busy designing a cover for my new book at the same time.

Since the barbershop her father usually goes to is closed because of the government's order, he has decided to have his daughter cut his hair. That is to say, to use her father's hair to practice her haircutting skill.

On Mother's Day, my son came to see me wearing a facial mask and asked me, what I thought of his hair?

I, as a proud grandmother, what else could I say? I said to him: "Her haircutting skill will improve with more practicing. Look at it this way, you have got the cutest, most beautiful barber in the whole world!" My son, the doting father, immediately laughed out loud, nodded his head nonstop, agreeing with my statement: "Yes, mom, you are right. She is indeed the world's cutest, most beautiful barber!"

世界上最美麗的理髮師

　　這張照片是我孫女兒四個月的時候跟她的小朋友們一起照的。一批小寶寶都是四個月大小，所以還都不會自己坐直，就算被大人們排起來坐在那裡也都東倒西歪，其中那個由左算第二名的那名坐得最不好，穿了那套大姑姑替她手織的粉紅色毛衣的小嬰兒，就是我的孫女兒。

　　女大十八變，應在我的孫女兒身上一點也不錯，長大了以後的孫女兒，長得眉清回秀，有混血兒白淨細緻的皮膚，頭髮的顏色跟她的外婆一樣是赭紅色的。文章寫得好，可以替我修改我的英文文章，也會畫畫，家裡牆上掛滿了她畫的畫，有一次我到她們學校的大禮堂去觀劇，劇院的外面的牆上有一張從天花板到地板的巨大壁畫，佔了一整面牆，非常顯眼。我的媳婦笑著對我說：「奶奶，這幅畫畫家的中文名字叫做李維珍，就是妳的孫女兒呀！」

　　現在，防疫期間遵守政府守在家中的法令，大家最好不要外出，以免傳染。在舊金山醫學院做副系主任的兒子，每週只正式到醫院去一天，其他的日子都是守著手機在家中待命隨時應診。孫女兒大學畢業以後，在舊金山醫學院找到一份用電腦的工作，是一名正在受訓期間的練習生，要三個月以後才能轉正。她同時也忙着替我的新書設計封面。

　　既然孫女兒手巧，她父親平常去的理髮店關門，有事弟子服其勞，也就是說用他的父親的頭髮當作試驗品，在家中替父親理了一次頭髮。

　　母親節那一天，兒子帶了口罩來看我，問我他的頭髮理得怎麼樣？

　　你想，我這做祖母的當然只有得意的份，我對他說：「理髮的技術，自然是熟能生巧，日益進步的，難能可貴的是,替你服務的,是世界上最可愛、最美麗的理髮師呀！」兒子一聽，這位寵愛女兒寵到不行的父親 立刻得意地仰天哈哈大笑 不停的的點頭，同意我的說法：「是喔，替我服務的,果然是世界上最可愛、最美麗的理髮師哦！」

My theory of family happiness.

Edited by Sandra Chin

When my son was twenty years old, he brought home a very lovely 19 year old young girl and introduced her to his father and me. "Dad, Mom, she is my girlfriend and her name is Leah." This was the first girlfriend he ever brought home, and ten years later they got married.   So my daughter-in-law is also his last girlfriend and the only one we ever knew.

At present, both of them teach at the San Francisco Medical School , California. They have been married twenty-five years and are still very much in love. Therefore, it is very easy for me to be the mother-in-law. I stay one-sided, of course on her side!

For the Chinese new year festival, I bought a hundred percent silk kung fu outfit for my 15 year old grandson and asked my son, the father, to forward it to him. I casually mentioned that there are two kinds of silk cocoons. Those made by one silkworm, called a single palace cocoon, have only one string, are easy to spin, and simple to weave. But double palace cocoons are made together by two silkworms. People generally use these to make silk quilts because they are softer and warmer. Chinese love to use these quilts as wedding gifts for they are auspicious and romantic.

Upon hearing this, my son said without hesitation, "Leah and I must be a double Palace cocoon!"

"Is that right?" I asked enthusiastically.

"Of course! There are thousands of silkworm cocoons in the same batch. What if I could not find Leah when we hatched out?" my son pointed out the problem very seriously.

Looking at my half-century old and still passionate son, my heart could not help but feel very, very moved. You can see that he is a very affectionate middle-aged man who treasures all of his blessings. I told him, "You and your wife discuss everything. As long as you have the same opinion, do as you wish."

"What if our opinions are different?" asked my son.

"Just do what she says. It's better then fighting and divorcing each other! Besides, weren't you promoted to be the director a long time ago? It's enough to be a master in the office!" Then I, the letting go type of mother-in-law, said to my son immediately, "I have changed my mind.   I will give this expensive kung fu outfit to my daughter-in-law and let her forward it to my grandson!"

"Why not me?" asked my son.

"Demonstration! I am showing you how to be one-sided (her-sided)!" I, the mother-in-law, answered with a big smile.

Then we, the crazy and silly mother and son, actually laughed very loudly.

作者與孫子

我的家庭幸福學

兒子在他卅歲時帶了一位十九歲白白胖胖極為可愛的小女同學回家，對他爸爸和我介紹說：「爸、媽，她的名字叫柯麗雅，是我的女朋友。」這是他帶回家的第一位女朋友，十年之後，兩人就結婚了，所以我的兒媳也是他最後也是唯一的女朋友。

目前，他們兩人都在加州舊金山醫學院教書，雖然已經結婚卅多年，可卻感情仍然很好，所以，我這個婆婆做起來就非常容易，只要一面倒就行了。

今年春節我替十六歲的小孫子買了一件百分之百的絲綢功夫上衣，本想交給兒子由他代轉。與兒子閒談中提起蠶寶寶結繭的時候，一般都是各自結繭，叫做單宮繭，只有一根絲，織起來方便又簡單，還有一种叫雙宮繭的，是由兩隻蠶寶寶共同結繭而成，一般用來做絲棉被，因為它真的比較鬆軟而暖和，中國人尤喜以之作结婚禮品，取其吉祥而充滿了羅曼蒂克的寓意。

兒子聽了，不假思索地應道：我和麗雅還是做双宮繭罷！

「是嗎？」我熱心地問。

「當然了！同一批结繭的蠶宝宝有成千上萬，萬一孵出來時找不到丽雅怎麼辦吶！？」兒子憂心憧憧十分認真地指出來。

看着己经年過半百"多情種子"的兒子，我的心中不由得不非常非常地感動，就笑嘻嘻地贊美道：「噯，可見你是一個十分專情而惜福的中年男子，所以，你們夫妻凡事應该協商，只要是意見相同了，都可以照你的意思做。」

「若意見不同呢？」兒子追問。

「就照她的意思做罷！一總比兩人吵架離婚好罷！何況，你不是早就升作主任了嗎？能在辦公室裡作主就夠了！」我這做婆婆的熄事寧人地下了结論，當時，我立刻對兒子說：「我改变了主意，還是把這件售價昂貴的功夫裝交給媳婦，由她轉交給孫子罷！」

「為什麼？由我轉交不好嗎？」兒子問。

「示範啊！教我兒子如何一面倒啊！」我這做婆婆的笑嘻嘻地回答。

我们這對又癡又傻的母子竟然相對大笑了起來！

2019 年 2 月 26 日世界家園

作者與孫子

The Bright Skylight

Edited by Adela Karliner(11/27/2020)

Crazy about the bright sunshine, we moved to Florida, the Sunny State, and built a waterfront house with a rectangular skylight on the entry hall's roof.

It added a lot of space and brightness to the room. Whether we were reading, watching TV, drinking tea or coffee alone, having a roomful of friends or a small group of intimates, it made us feel happy and comfortable.

On hot sunny days, the sun from the sloping ceiling down showed the cloudless blue vault of heaven through the skylight. A long square white bright light covered the floor. Inside the hall, the air conditioning quietly cooled off the room; it was another kind of laziness and relaxation.

Many indoor plants, such as orchids, ivy, lucky bamboo, and African violets thrive in this filtered light through the clear glass.

Since the sun keeps moving, the skylight croaks a very long piece of brightness. Putting some greens or colorful flowers on the shelves or tables where sunshine passes makes one's heart full of comfort.

At night, like misty waterfalls, the moonlight gently pours into the house, making the furniture visible. Standing on the floor looking up, one can see the beautiful moon. The moon's shape changes from crescent to full, then from complete round to a slit. Through the rectangular frame, the clouds sometimes the moonlight, and other times is as bright as a mirror. Sitting on a sofa, watching the rectangle-shaped light, and then looking out of the floor-to-ceiling glass window, one can see the road outside; silver screen appears to cover the grass and woods, making every tree, flower visible and beautiful if one is in a poetic dream. It is like the wonderland of a fairytale.

On a rainy day, the sound of raindrops hitting the skylight is an indicator. If it pours urgently, it is a heavy rain. When the sound flows down gently, it indicates that it is lightly raining. When it seems as if there is no sound, then it is drizzling outside. One feels the bed as warm and cozy in the bedroom, hears the music from heaven, ding dong ding dong, and the lullaby gently puts us into sweet dreams.

One day, with a pacifier in his mouth, our young grandson crawled on the floor, chasing the shadow formed by the skylight. He was giggling

all the time.

His clever sister said to me, "Grandma, the sun has feet, its shadow moves!" What she said took me aback. Wasn't that the eminent writer Zhu's famous saying? Why did she think it when she was only five years old?

Suddenly, I realized that the saying was not the celebrated author's most valuable possession, but the pureness and innocence of a newborn's heart?

Peony

明亮的天窗

我們因為喜愛太陽，所以退休以後就搬到名叫陽光州 Sunshine State 的佛羅里達，造房子時還特地要建築師在大廳的屋頂上設計了一個長方形的天窗。

有了它，廳內增加了不少寬敞和明亮，在裡面無論 讀書、看電視、飲茶、喝咖啡，或高朋滿座、或知己小聚都覺得心曠神怡，非常舒暢。甚至炎熱的艷陽天，蔚藍天空裡萬里無雲，太陽由天花板斜斜地照射下來，地板上鋪著長長方方、白花花的一大片亮光，室內冷氣閒閒地、不動聲色地開著，也另有一番慵懶舒散的風情。

很多室內植物如蘭花、常春藤、富貴竹、菲洲紫蘿蘭等都喜歡這種玻璃透過來的弱光，又因為太陽不停地移動，所以由天窗上照過的地方有很長一片，在適宜的架上、几面我都放上一些翠綠艷紅，只要看著它們，心裡就充滿了知足快樂。

夜晚的月光，似水一般地瀉入屋內，照得廳內傢俱桌椅都看得清清楚楚，站在地板仰首向上看，天窗內美麗的月亮，時圓、時缺，有時晦暗，更有時亮如明鏡，坐在印著長方形光亮的沙發上，癡癡地朝窗外觀看，外面的道路、草地以及樹林裡，每一株樹、每朵花都披著銀色的細紗，清晰而美麗，如夢、似詩、更好像在仙境裡一般。

外面下雨了，天窗上傳來淅淅瀝瀝的雨聲。若窗上映出的雨水嘩嘩地急傾而下，那就是天降大雨，雨水緩緩地向下慢流，表示外面正在下小雨，若只有毛毛雨的話，玻璃上只不過結了層薄霧而已。在臥室睡覺，被窩裡溫溫暖暖，耳中隱隱傳來天籟之聲，叮叮東東，自然輕柔地引人進入甜蜜好夢。

一天，一歲憨憨的小孫子含著奶嘴，爬在地板上追逐著由天窗照過來的太陽影子玩耍，傻笑之聲不斷。

聰明的小姐姐對我說：“奶奶，太陽是有腳的，它的影子會移動的呵。”我吃了一驚，這不大文豪朱自清說過的話嗎？怎麼才五歲的她也這麼想呢？

突然領悟到，原來文人最寶貴的擁有，不就是一顆赤子之心嗎？

One Dollar Rental

Edited by Sandra Chin

Since I sold my Florida house at the end of the year, I have been staying temporarily at my son's house in San Francisco, California. My son, daughter-in-law, granddaughter, and grandson are all very sweet to me. We have a party almost everyweek, so it is very hard for me to tell whether they are celebrating my moving west or it just coincides with the festival season. But, does it matter for what reason so long as everybody is cheerful?

The main thing is that I should start looking for a permanent place for to live. However,I am shocked to find that in cold and misty San Francisco, every place is so expensive!

One day, I get off the phone with a friend who is also a realtor. I am going to walk to the kitchen to brew myself a cup of coffee when my daughter-in-law comes over.

"Grandma, don't you like living here?" she asks sweetly.

"Of course I do, here is great!" I reply with a grin. To be honest, I have no complaints about the room where I am staying. It is spacious with a high ceiling. A curtain separates the comfortable sleeping area and the work area which contains a bookshelf and a small desk with a computer and phone. It is simple, convenient and perfect for my needs. There is a balcony outside the glass window and beyond that there is a courtyard. Looking through the floor length window, I view beautiful downtown San Francisco.

"Then, Grandma, why are you in such a hurry to look for an apartment?" she asks.

"Hee hee, have you ever heard of the celebrity, perhaps Mark Twain, who said that relatives are like dead fish? Living in the same house for too long, there will be a rotting smell!" I answer with a smile. It is better to tell the truth, of course, and at this moment, openness is the best policy.

"Is that right? What if you pay rent; will that make you feel better?" she asks.

"Uh…." I don't know how to answer.

"What about you come up with one dollar as the rent?"

she suggests with a smile.

"One dollar as the rent?" I still don't get it.

"Yes, Grandma NeiNei. Do you remember that you once sold us a car for one dollar?" my daughter-in-law reminds me.

Hearing her suggestion, I immediately remember that a long time ago, my son brought his girlfriend home to visit us. At the time, they were both students in medical school in San Francisco. We found out that they had no car and it was very inconvenient for the two young love birds. So we sold them our car for one dollar. We never dreamt that twenty years later my son's girlfriend would be our daughter-in-law, the mother of our beautiful eighteen year old granddaughter and handsome fifteen year old grandson, and that I would be living temporarily in their house. I cannot help but sigh at how fast time has passed, and how things have changed so much!

But now it seems to me that until I can find a suitable place, I will happily accept my daughter-in-law's proposal to rent my room for $1!

一塊錢美金的房租

　　自從賣掉佛州的房子，搬到加州以來，正值歲末佳節，一直跟了兒子、媳婦、孫女及孫兒，天天赴宴，日日笙歌，兒子李德康還當眾宣布，為了慶祝老母西遷，也要在自己家中大肆請客，表示歡迎之誠意，到目前為止，完全一片母慈子孝，前途大好的光明景象。

　　本老太太每天吃香喝辣，除了努力維持自己慈母形象之外，也有一絲自覺，認為是該考慮長住久安之計的時候到了，就去打聽了一下，不問不知道，一問嚇一跳，原來整日價淒風苦雨的舊金山，不但米珠薪貴，房價也高得離譜，就私自尋思，打算在三月底由台灣回來之後，在風和日麗的南灣暫租一處價錢公道合理，房租在自己能力範圍內的公寓，以便自由自在地享受餘生。

　　如此，我就到處忙碌打聽居處情況，大概動靜過大罷，驚動了我那猶太裔的美籍兒媳。

　　這天，我打完電話給朋友及房屋經紀人，正放下電話，打算站起來走到廚房中去煮杯咖啡給自己喝的時候，兒媳走了過來。

　　「奶奶，妳不喜歡住在這裏嗎？」她和顏悅色地問。

　　「沒有不喜歡，這裏太好了！」我笑嘻嘻地回答。說實話，我暫住的這房間實在沒有什麼好挑剔的，寬敞的房間，高高的天花板，睡覺的床邊有一個書架及窗簾隔開外面小書桌，在桌上寫作，計算機、電話都十分完善方便，桌邊明亮的落地窗外還有一個不大不小的陽台及院落，隔窗放眼望去，舊金山市區美景一覽無遺。

　　「奶奶，那妳為什麼這麼忙著找房子呢？」她又問。

　　「嘻嘻，妳沒有聽過一位名人，可能是馬克吐溫罷，好像這麼說過，大意是親戚和死魚一樣，在同一房屋內居住過久，就會有腐爛的氣息？」我笑嘻嘻地據實以告，此時此刻當然開誠布公比較好。

　　「是嗎？妳如果付我們房租，是否會覺得心中好受些呢？」她追問。

　　「‥‥」我不知如何回答。

　　「那妳拿出一塊錢美金給我們當房租好了！」她笑嘻嘻地建議。

　　「一塊錢美金當房租？」我還沒有弄懂。

　　「是啊，奶奶，妳記不記得妳曾經向我們收了一塊錢，把一輛汽車賣給我們？」媳婦提醒我。

被她這麼一說，我想起來了，那是早先我兒子尚在讀舊金山加州醫學院的時候，帶了他當時同居的女朋友回家看望父母，我們看他們兩位年青學生沒有汽車，生活起居極不方便，就把我們正在開的車送給他們代步，在汽車轉戶的欄上填寫的是他們以一塊錢美金的代價來向我們購買的，沒有想到事隔廿多年，兒子當時的女朋友，不但變成了我們現在的兒媳，也變成了十八妙齡美麗的孫女兒及十五歲高大英俊孫子的母親，而我目前現在正住在他們家暫時過著依親生活。

此事此刻，實在令人忍不住要唏噓歎息，時間飛馳，世事變化得真快呀！看來，在沒有找到合適的房屋以前，只好接納兒媳的建議了！

My Flawless Son

Edited by Descartes Li

Today was a bit special. It is a family get-together day; all the men went out to the bar to watch super bowl. All the women stayed home. My daughter-in-law brought a cup of steamy coffee to me.

"Nei Nei! here is   a cup of hot coffee for you!" She smiled and put the hot coffee cup in front of me, then sat down beside me   and looked as if she had something to tell me.

" You don't have to go to work today!" I said, as I drank the hot coffee.

"Today is national holiday. Nei Nei, what do you think of your son Kang Kang?" she asked. For easy separation, the entire family called me Nei Nei; a Chiese name for paternal grandmother. As for their maternal grandmother, they use English grandma to distinguish these two, because in English, there is no difference between these two grandmothers.

"What do I think of my son   Kang Kang? He is a perfect good man with zero defects! He was quiet and well-behaved little boy when he was young, and I never have to be bothered to check his school work nor his outfit! Sometimes in the early morning of the cold winter, he even turn on the heater in the car to warm up the engine as well as the air inside of the car then he would sweep the snow on the car roof!" I praised enthusiastically.

"One shouldn't wake him up while he is napping! He gets up early in the morning and works so hard! He must   take a nap at noon time." I quickly explained before she open her mouth.

"well, any thing was already planned. Changing it may involve too much." I defended my son whole heartedly without his wife uttering any word.

"Mom, didn't I say so? It is the same! In Nei Nei's eye, her son is just like our flawless dad in the eyes of my grandmother!" my daughter-in-law turned her head facing her mother, an American born Russian lady.

"Oh, it's like this. My husband is one hundred percent perfect in the eyes of his mother, my mother-in-law." My daughter-in-law's mother smiled and added. My in-law's mother-in-law, was an old English lady who had owned the largest local grocery store in a town in New Jersey, and later sold her store and moved to California to be near their only son's family, and taking care of their grandchildren.

"Is it so!" I answer.

"Truely, my English mother-in-law thinks her son needs no room for improvement!" my daughter in law's mother couldn't help laughing.

"Howerever, my mother thinks differently!" Her daughter, my daughter-in-law also laughed.

At this time, my grandson showed up in sleepy eyes, and his grandma smiled and asked, "Sleep late?"

His mother, my daughter in law who was a physician, got up quickly and said to her mother, "Mom, He slept late last night and the child had to sleep well to be healthy!"

"Of course! All mothers in the world think that their sons are flawless! There is a five thousand old Chinese saying; the only good man is your own son!" I laughed and shouted out loud.

"Yes! All the sons are flawless, while Husbands have countless problems that need to be improved!" The three women who had made mother-in-law, mother, and grandmother laughed and claimed in unison.

我的零缺點的愛人

這天有點特別，是家人團聚的日子。所有的男人都到酒吧去觀看足球總決賽，女人們全待在家。我的媳婦手中端了一杯熱氣騰騰的咖啡，踱到我的房間來。

「奶奶，請喝熱咖啡！」她笑嘻嘻地把手中的熱咖啡放在我的書桌上，自己一屁股坐在我的床邊，看樣子有話要跟我講。

「嘿，今天沒有去上班？」我問。

「今天全國放假。奶奶，您老人家覺得您的兒子康康怎麼樣？」她問，為了容易分別起見，他們全家都用中文叫我奶奶。至於他們的外婆呢，就用英文的祖母，以資區別，因為在英語中，內祖母與外婆是沒有什麼分別的。

「妳說康兒嗎？他是個零缺點的兒子喔！從小就安靜乖巧，書包衣服都不要人煩心，甚至連寒冬的清晨，他一早就把自己整理好，坐到車上，先把汽車引擎打開，讓暖氣充滿了汽車，又把車頂上的積雪掃乾淨……！」我非常熱心地嚷道。

「本來就不應該吵醒他嘛！他每天一大早就起床，工作那麼忙，只有中午睡午覺時休息一下……。」不等媳婦開口，我連忙解釋。

「是喔，早就計劃好的，要改一下，大概牽涉過多吧……。」我辯解道。

「媽媽，妳看我說的不錯吧！康康在奶奶的眼裡，就跟爸爸在我祖母的心　頭一個樣子，是沒有缺點的。」媳婦轉過頭來對跟在她後面的母親說，媳婦的媽當然就是我的親家母。這位在美國出生的俄國老太太什麼時候跟在她美國女兒的身後，也玉駕親臨了我的房間呢？

「嘻嘻，是這樣的啦，我的老公在他母親的眼也就是我婆婆的眼中是百分之一百的完整人物……。」親家母笑著接口。親家母的婆婆，也就是親家翁的母親，是一位英國老太太，以前在新澤西州的一個鎮上開了一家當地最大的雜貨店，後來後來為了獨子的事業，出售了自己的商店，特地搬到加州來照顧孫子、孫女兒。

「是嗎？」我問。

「真是，我的英國婆婆居然覺得她的兒子竟然是沒有缺點的！」我的親家母忍不住笑了起來。

「可是我媽媽覺得我爸爸要改進的地方多了去了！」我那做

醫師的媳婦笑了起來。

「不錯！世界上的媽媽覺得自己的兒子是一位零缺點的愛人！也就是我們中國人說的叫做：兒子是自己的好！」我大聲地下了結論。

「不錯不錯，丈夫們卻是急待改進！」三位做了婆婆、媳婦、祖母的女人都哈哈大笑起來「不錯不錯，還是有五千年歷史文化的中國人說的話對，兒子是自己的好！」

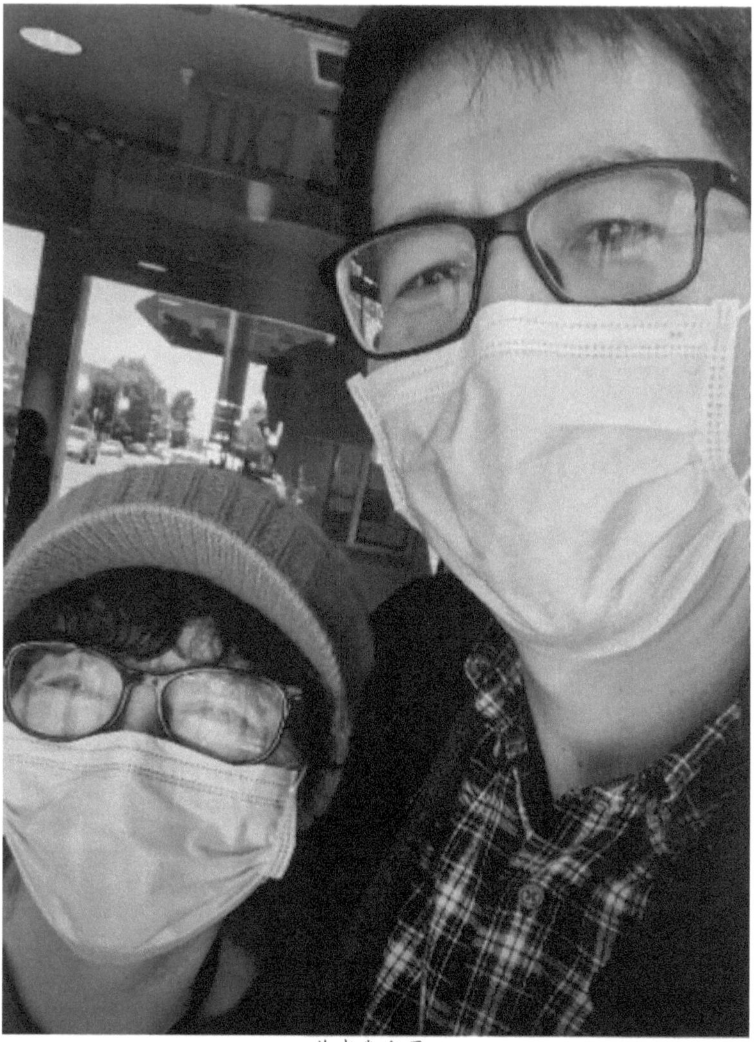

作者與兒子

Edited by Adela Karliner (08/23/2020)

This morning Grandfather Li gets up early, and happily goes out for a walk to visit his friends as usual. On his way back, he stops at the newsstand at the street corner to buy today's newspaper. He is surprised by the increase in price, so he foregoes eating his usual baked cake, fried stick, and soy milk, and comes home with an empty stomach.

"You see, today's newspaper has only four pages, but costs $3.50, it is really not worth my money," our old gentleman says, very dissatisfied, as he eats his white rice porridge with salted tofu at home.

"That's because the newspaper has canceled all commercial advertising, and the printing costs are distributed among the readers," his son says while drinking his coffee. The son uses his hand to check the weight of the newspapers.

"The thin one is more expensive than the thick one. The world is definitely getting worse by the day!" after drinking porridge, Grandpa Li says as he wipes his mouth indignantly. Then he decides to take the bus to the cultural and educational center to play chess with his old friends.

"Cutting down the size of the newspaper saves us a lot of time. Two-thirds of newspapers are usually advertisement, which is a total waste." After drinking his coffee, the younger Mr. Li takes his briefcase and hurries out to work.

"What's the matter? Where are the advertisement pages in today's newspaper? Which beauty salon has a discount sale on permanents? What shall I do?"   Mrs. Li asks.

"Alas! I am miserable. I have invited baby Zhu to go to the movies. After watching the movie, we plan to go to McDonald's to eat hamburgers. Now, it is 5:30 PM, I do not know what movie theater is playing which movie, and do not know why the newspaper does not have the 'buy one get one free' coupon. I only have $5.00 in my pocket. Baby Zhu will arrive soon, what shall I do?" Big Brother Li scratched his ears and was very fidgety.

Mother Mrs. Li fluffed her messy hair and was unhappy for a whole day. The simmered meat on the table was a little charred, and the stir-fried green vegetables turned yellowish.

The whole family was somewhat grumpy. When everyone was seated for dinner, they found that Little brother Li was missing. Suddenly they all hear someone in the basement crying so loudly that it shakes the wall. It turns out that the puppy training class has trained Little Brother Li's dog to use the newspaper ads as toilet paper. It only recognizes those pages and is now rejecting any substitutes. So today, the basement is totally disordered.

After the meal, Mr. Li said, "Hey, I just don't care, as long as the television still has ads, I'll have some time to go to the fridge to get a beer and go to the toilet if I need to!"

When Mr. Li is about to sit on the sofa to relax, suddenly, he hears his wife shouting from the bathroom as if there were a hundred thousand things on fire: "The father of my children, the toilet is blocked! There are no ads in today's newspaper, I can't find the phone number for the Guaranteed-Through Pao. Quickly drive to the Guaranteed-Through Pao's home. You must be fast, the toilet water is going to overflow to the whole bathroom!"

報紙今天沒有廣告

李家爺爺今天清晨與往常一樣，一大早起來歡歡喜喜地出門散步訪友，回來時在街角的報攤上順便買了一份當天的日報回家，不意報紙漲了價，所以沒有捨得在外面吃燒餅夾油條配豆漿，餓著肚子回家。

「你們看，今天報紙才四張紙，要我$3.5毛錢，真是划不來。」今天他老人家在家裡喝白粥配豆腐乳，一面很不滿意地說。

「那是因為今天的報紙取消了廣告，報紙所有的印刷發行的成本都要我們這些看報紙的人分攤的緣故。」李家爸爸正在喝咖啡，聽見爺爺埋怨，就順手將那幾張報紙取過來惦了惦。

「薄的反而比厚的貴，這世道真是一天不如一天了！」喝完了粥，擦了擦嘴，李家爺爺很憤慨地說，決定還是安步當車到文教中心與老朋友們下棋去了。

「節省了很多時間，平常報紙三分之二都是廣告，等於浪費。」李爸爸喝完咖啡，提了公事包匆匆出門上班去了。

「咦？今天報紙的廣告頁到那裡去了？我要看哪家美容院燙髮打折扣？如何是好？」李家媽媽大驚小怪地問。

「唉！我真慘，說好了要請朱寶寶去看電影，看完去麥當勞吃漢堡包，現在已經5:30 PM，還不知哪家電影院演什麼，又不知道為什麼報紙上沒有買一送一的禮卷，口袋裡只有五塊錢，朱寶寶馬上就要到了，把我給急死了！」李大哥急得抓耳撓腮，坐立不安。

李家媽媽蓬了一天的頭，心　一肚子氣，端上桌的紅燒肉有點焦，炒的青菜也發黃了。

全家個個滿肚子委屈，坐上飯桌，數數少了一個李小弟，要去喊他，突然聽見地下室有人哇哇大哭，聲震屋瓦。原來李小弟的小狗狗在訓練班受訓時都是用報紙的廣告當作入廁用紙，它只認得那幾版，現在拒絕任何代替品，所以今天地下室滿目瘡痍，不堪入目。

「嘿，只有我不在乎，只要電視還有廣告，我就有空檔到冰箱去取啤酒，到廁所去方便！」飯後，李家爸爸由廁所出來，正要朝沙發上坐，突然聽見李媽媽喊叫之聲，她十萬火急的大聲喊道：「孩子們的爸爸，馬桶堵住了，今天沒有廣告，報紙上找不到包包通的電話號碼，你火速開車到包包通家去一趟，要快一點，馬桶的水已經漫滿啦整個廁所啦！」

I love You, Adela

Edited by Adela Karliner (08/27/2020)

Adela is a beautiful, intelligent lady with a kind heart. She loves all her children as well as her children-in-law, and one of her sons-in-law is my son. So Adela is related to me by our children's marriage.   There is a Chinese name for our relationship called "Qin Jia mu."   In ancient times males and females alike all wore clothes with long and roomy sleeves. That also meant our children could form one family, and we could share the same group of grandchildren. So, the two of us are qualified to tie our sleeves together, and our hands inside the sleeves, can complement each other, and help each other, planning secret events unknown to outsiders.

My in-law Adela and I did have the chance to plan events together several times. But until today, I have been relying on her to help me get through my difficulties. It, of course, is a good thing for her. She does not need my help. But from my point of view, I do wish that I could have reciprocal.

Now let me recall what she did to help me!

The first thing she did to help me was when my husband had to retire early because of ill health, had to concentrate on rest and recuperation. At the time, because the company I worked for had some financial issues, and tried to encourage their employees to retire. Adela used her lawyer's position and knowledge to help me with legal procedures to secure my future life, so I could retire early with my husband at the age of 53 and move to Florida. He happily made friends fishing and maintaining his health in our waterfront two-story house. I could dedicate myself to writing.   Thanks to Adela for her help.

After sixteen happy years, my husband passed away. We had known each other for close to 58 years.   I did not want to stay alone in Florida. So again, I asked Adela for help.   At the time, she also had retired. She introduced her other in-law Michael Chase, a lawyer familiar with Florida law, to help me with all my financial arrangements and real estate transfer matters, and it was done free of charge. That was so much needed for me because I could not afford expensive lawyers' fees.   It was the main reason that I could leave Florida within a year and adapted to a new single senior life.   I was grateful and appreciated Adela and Michael's help very much.

At present, we are staying at home following the government COVID-19 pandemic orders. I am self-publishing a collection of my short stories with English translation. Because English is not my first language, I have asked all my highly-educated nieces and nephews to review my English. Although their English training is proper, but they all are relatively young. A book always needs people with seniority, more famous, experienced ones to make the Grand final edit. Adela, who had taught English classes at San Diago State University, was the best choice for me. Putting her name at the top of our editor's list and posting her picture where everyone can see would naturally push our book several notches up. And, I am sure that it would please my readers a great deal.

Now, one of the things I need to do for the rest of my life is to find an opportunity to repay my sleeve connection.

For Adela, I must find some opportunities, with practical action to show that I love her, admire her, and respect her very much!

Author and Adela

親家母柯艾達

　　柯艾達長得漂亮，聰明，加以有一顆善良的心。她的丈夫曾任美國心臟醫師協会的主席，她愛她的家人,愛她的女兒，也愛她的女婿。她的女婿是我的兒子。所以柯艾達就是我的親家母。古時候中國男女的袖子都是又長又大，既然我們兒女可以成為一家人，我倆共享共有孫子及孫女。所以，我們有資格把衣服袖子連在一起，兩人的手可以在袖子裡面，互補互助，互相幫忙，策劃一些不為外人所知的秘密事件。

　　我跟我的親家母柯艾達雖然也籌劃了一些事情，但直到如今為止，都是靠她幫了我的忙來度過難關。她尚不需要我幫忙，對她雖然是好事，对我這一直想回饋她的人來說，似乎有點遺憾。

　　現在我們來回憶講述一下，她幫過我什麼忙吧！

　　她幫我忙的第一件事情，就是我的先生因為身體欠佳的緣故，必須提早退休專心休養，正好我工作的單位，因公司的財務的關係，也盡力鼓勵我們退休，我的親家母柯艾達知道了此事，立即用她律師的身分和知識，幫我辦理合法手續，以免後患，所以我可以在 53 歲的時候就與我先生提早退休，搬到佛羅里達去了。在水邊我們自建的两層樓房裡，他快樂地捕魚交友，保養身體。我則專心一致寫作，非常感謝我們親家母的幫助。

　　十六年後，結婚多年的老伴過世，我不願意自己一人再留在佛羅里達，以免覩物傷情，那時我的親家母也退休了，不过她又出面幫我找了她另外女兒的公公，一位熟悉佛州法令的親家律師蔡麥可免費幫助我辦理一切未亡人要辦理的動產和不動產的轉移事項，我能夠盡快地辦好一切法律手續，早一些搬到加州，適應單身老人的新生活，無憂無慮过著靜好的歲月，真是必需感謝親家母柯艾達了。

　　我想在 2020 年出的新書加上自己的英語翻譯，但是因為英語不是我的第一語言，所以找了受過高等教育的姪子姪女們把我的英文編審一下，她們英語修養雖然不錯,但是到底比較年輕，總得要有一位輩份較高,名氣較大、經驗比較豐富的人來壓軸才是，捨她其誰，當然是在聖地亞哥州立大學做過英語教授的親家母柯艾達名氣最大,經驗最豐富啦。把她的名字放在我們的編審首位，把她的照片放在大家都看得見的地方，就自然會把我們出的書增加很多檔次，一定更受讀者们喜愛囉。

　　現在，我的餘生需要做的事情，就是尋找機會能夠報答我的親家母艾達，對於她，單單用言語告訴她說我愛她是不夠的。她今年的生日快到了，我想利用大家到每年都去的渡假屋的機會，鹵製一些中國鹵味給大家一同食用，希望能表示一下，我愛她，佩服她，而且尊敬她的萬分之一！

作者與柯艾達

I love Rainy Days

Edited by Adela Karliner(12/28/2020)

Dark clouds gradually covered the hot sun, but the air was getting more and more sultry. Pulling weeds in the yard, I was sweating. I tried to breathe hard through my hay-fever-stuffed nose. When I was about to turn on the tap water to help the wild grass fighting against the scorching sun all day long, the sky began to thunder a dull loud sound. A large cloud from the east gradually moved in our direction. It first scattered a little bit of water and then quickly spilled more significant drops in our area.

Great, the rain was coming. A senior lady like myself didn't have to continue my hard work! God was going to water the flowers for me! Before I retreated to my home, I heard the joyful sounds of children playing in the backyard river. So I shouted, "Little friends, it's raining. Let's all go home!" Probably God wanted to increase the weight of my speech, suddenly sounded a few more big peals of thunder followed by rumblings. And, the lightning ring lit up half the dark sky, those little players in the river saw the ferocious flashes and were scared by its loud sound. One by one, they jumped out of the river and ran home wet.

After I made sure that all the little ones had dispersed, I hurriedly pack up my gardening tools. In the light rain, I quickly dragged my small tool cart home.

Opening the door, I found the house's air blowing in my face was nice and cool. After I bathed, I changed to clean clothes, sipped a cup of hot tea, sat down on the sofa, and heard the skylight hit by the wind and rain. It was so pleasant to listen to the tranquil sound of raindrops on the roof. The raindrops' light tone gradually grew louder from the skylight. The jingling music evolved into a roaring pour. Looking out of the window, I saw the wind was helping the rain's power. The coconut leaves usual leisurely swayed in the wind. Now, they were turning left to right, struggling in the harsh blowing, striving for survival.

After a while, the downpour of heavy rain slowly lighter and gradually stopped. From the drainage pipe along the eaves of the house kept flowing a small amount of clear water. The trees still dropped some raindrops.

I pushed the door open and found croaking frogs surrounding the backyard. Ah, the small pool outside the house had become a bigger pool from the rising water. Heavy rain had hatched a pond full of lively small fish. They were chasing and playing with each other in the water. An elegant and beautiful white stork with long legs quietly stood in the middle of the pool. Was it waiting for the small fish to grow up?

The air was so fresh that my hay fever was completely gone! The sun re-opened it's smiling face. The trees and grass had all turned emerald green, the flowers were more colorful than before, and the rain-washed cement driveway was immaculate. A large turtle crawled out of the river, and its' four feet moved very firmly forward.

Squeaking, squeaking, two teenagers, each rowing a small leather boat, flew along the river. Hu! Hu! My old man was practicing throwing his newly bought fishnet into the water. Ha! Ha! Flocks of blackfish jumping in the water, laughing at the retired old professor, because the fishnet he threw out was not the correct circular shape.

My mind was still on the giant turtle. I was afraid that the passing cars might crash into it. When I decided to help the slow creature, it had already crossed the road and emerged onto the other side's grass.

我愛下雨天

　　豔陽漸漸被烏雲蓋住，空氣愈來愈悶熱，在院子裡　拔了一些雜草，我滿身大汗，鼻子被花粉熱塞住，呼吸　不順暢，正要打開自來水開關，給那些終日在大太陽下　與野草抗爭的家花澆點清水，遠遠的天空中開始響起一　兩聲悶雷，眼見一大片烏雲由東面漸漸向我們住的方向　移動，再過了一下，點點滴滴，東邊天上的一批水珠，飛快地向我們這邊灑將過來。

　　好極啦，下雨嘍，不勞我老人家再繼續辛苦，老天爺開始替我灑水澆花了！當我正要向家中撤退時，聽見後院河中仍有兒童戲水之聲，連忙大聲喊道：「小朋友們，天下雨了，快些回家去罷！」大概老天爺要增加我說話的份量罷，突然間打了幾個大雷，隆隆的巨響中夾著閃電，照亮了半個烏黑的天邊，河中的洋娃娃們看見雷電來勢凶凶，全部被嚇得大呼小叫，一一由河中躍起，濕淋淋地冒雨狂奔回家。

　　我見他們各作鳥獸散去，也連忙收拾種花的工具，在小雨中匆匆拖了小小的工具車，向家中跑去。

　　打開大門，屋內的冷氣迎面吹來，真是涼爽，洗過澡，換上乾淨衣服，泡上一杯熱茶，在沙發上坐下，仔細欣賞天窗上傳過來的風聲雨聲，真是聲聲入耳，令人心曠神怡。點點滴滴的小雨落在屋頂上，由天窗傳來淅淅瀝瀝的輕響，雨滴漸下漸大，叮叮咚咚的聲音演變成嘩嘩啦啦的傾注，移目朝窗外看去，只見風助雨勢，平日悠閒迎風搖曳的椰子樹葉，正在大風雨內左搖右晃，努力掙扎，奮勇求生。

　　傾盆的大雨慢慢變小而逐漸停止，朝外再看，屋沿的排水管尚流淌著一些清水，大樹下面還答答地滴下一些水珠。推開大門，後院四面八方都是蛙聲，嘓嘓地響成一片，啊，原來屋外的小池漲水，大雨已經孵出一池活活潑潑的小魚，在水中追逐嬉戲，一隻優雅美麗的白鷺，長腿沒入水中，靜靜地站在的池水中央，是在等待小魚們長大些嗎？空氣被洗滌得如此清新，花粉熱早就不見了！太陽又重新展開笑臉，樹木特別青翠，草兒格外碧綠，花朵兒比原先更加嬌豔，水泥車道也被清洗得乾淨清爽。一隻很大的烏龜由河邊亂石中爬出來，四隻腳非常堅毅地向前移動。

　　吱，吱，兩名青少年各劃了一艘小皮舟，沿河飛駛而過。呼！呼！原來我家老爺子又在用他新買來的魚網在練習撒網，成群的烏魚在水中跳躍，哈，哈，它們在嘲笑這位退休的老教授，丟出去的魚網，一直撒不成圓形罷。

　　我心中惦記著，怕那只大烏龜被路上行駛的汽車壓到，轉過身打算助它一把，但它早已穿越馬路，沒入對面那家人家的草叢中去了。

Remembering A Lover from a Previous Life

Edited by Descartes Li

When we were young, our family lived in the distant outskirts of Chongqing, China. Drinking water was collected from a pool outside the school by a school janitor. After treating the water in the small tank with alum, he shouldered a long bamboo pole to carry the water from the pool to the water tank of each teacher's house. The water was so precious that we never used it to take a bath. In addition, as it was not easy to heat the cold water either, our parents took me and my brother to a small town called South Hot Spring to take a warm sanitary sulfur bath.

One morning, when it was still dark in the early morning, my father held my small hand and took me to walk on foot. My mother and my toddler brothersat in a hired a bamboo sedan chair carried by two sedan bearers, along with a home cooked meal. Although the we set off at the same time, I was little and could not walk very fast. The sedan bearers walked very fast, so before long, the two groups of people were farther and farther away from each other, and soon my mother and brother, gliding along in the sedan, disappeared.

I tried my best to hurry my little short legs in the darkness. We saw a faint light in front of us on the roadside. As we got closer, we discovered that the light came from a oil lamp hung on a soup stall. Dad finally realized that I was running so he slowed down his pace.

"Gwen, are you hungry? Do you want to eat a bowl of Yangchun soup noodles?" Dad lowered his head and kindly asked me.

"Daddy, I'm not hungry!" I know that my mother's had cooked all day long at home the day before, and the homemade food is wrapped in her luggage. If I eat the noodle soup on the roadside stall, we will have to spend money. I wanted to be a good child so I didn't want to spend my father's money.

"Gwen, are you tired? Do you want me to carry you?" Daddy asked me kindly.

"Daddy, I'm not tired, you don't have to carry me!" I replied loudly. How can a good girl let herself be carried by her father?

After walking for a while, Dad finally said, "Gwen, how about let me carry you on my back so we can walk faster and catch up with your mother and brother sooner?" With that, Daddy knelt down and let me climb up on his back.

By that time, the sky was already a little brighter than before, and I could vaguely see the shadows of the vegetables growing on both sides of the dirt road. I held on to my father's back and felt that his back was wide, warm and big. I felt so safe and so happy.

Dad was always so very fond of me, and I of him. Ah, the memories of the two of us sitting and singing together on a bench in our back yard or telling stories to each other as we lay on the straw mat in the middle of the house.

There is a Chinese saying: "The daughter is her father's lover in a previous life." While a man's favorite should be his wife, their daughter is the crystallization of their love, a combination of his wife's family progenitors plus his own cherished ancestors. How many generations of wooden fish are knocked to receive this good fortune?

Being able to be my father's daughter in this life was lucky enough. And the sweet young man who later became my loving husband was not exactly like my father, yet was another bounty of good fortune.

But that is another story.

我是他前世的情人

　　小的時候我們家住在重慶市郊巴縣界石鄉，食用水是國立邊疆學校的校工將校外小池子裡的水用明礬處理過後，再用扁擔挑到每位老師家的水缸裡，吃的水這樣金貴，那洗起澡來一定也不很暢快，加以將冷水燒熱十分不容易，所以每月一次，父母就會帶了我和弟弟到旁邊一個叫南溫泉的小鎮去好好洗個暖和衛生的硫磺澡。

　　那天天不亮，爸爸牽了我的小手帶我徒步行走，媽媽帶了初生還不會走路的弟弟坐滑桿轎子帶了食物及換洗的細軟前往，雖然兩起人一同啟程，可惜我尚年幼，走路不夠迅速，而轎夫們健步如飛，所以後來媽媽和弟弟坐的轎子與步行的父親和我相距越來越遠，不久就不見了他們的蹤影。

　　我努力邁開我小小的短腿，在黑暗中盡量跟著跨著大步的爸爸向前奔跑，在路上遠遠就見到前方有一絲微弱的光亮，走了好一陣子，靠近了才發現那亮光是由一個湯麵攤子旁的桐油燈發出來的。爸爸終於覺察我在奔跑，格外放慢了腳步。

　　「國英 妳餓了嗎？要不要吃一碗陽春麵？」爸爸低下頭來，和善地問我。

　　「爸爸，我不餓！」我知道媽媽包袱中帶了一些昨天忙碌了一整天在自家煮好的食品，臨時吃路邊麵攤上的湯麵是要另外花錢的，好孩子不要讓爸爸亂花錢，我要做一個好孩子。

　　「國英，妳走了這麼遠，累了嗎，要不要爸爸抱呢？」爸爸很慈祥也問我。

　　「爸爸，我不累，不要爸爸抱！」我大聲地回答，好孩子怎麼可以要爸爸抱呢？

　　又走了一陣子，爸爸終於說：「國英，還是讓爸爸揹著走罷，這樣我們可以走得快一些，早點與媽媽和弟弟在一起。」

　　爸爸蹲了下來，讓我趴在他的背上。

　　那時天已經快要亮了，可以隱隱約約地看見土路兩邊種的蔬菜的影子，我伏在爸爸的背上，覺得他的背又寬又大又暖和，自己既安全而又幸福，非常非常之喜歡我的爸爸。

　　終此一生，爸爸非常寵愛我，而我也更是一直都非常非常的喜歡我的爸爸。我們父女可以一同坐在院子裡的長椅上唱歌,也可以躺在屋內蹋蹋米上一齊說故事。

　　長大了之後，聽見中國有句話說：「女兒是父親上輩子的情人。」覺得真是非常之中肯，父親和女兒之間是有一種很特殊的情愫的。

　　是什麼樣的情愫呢?原來這種父親女兒之間的情愫要用現代的遺傳科學來解釋才能更加明晰,因為男人一生最愛的就是他牽手娶到家的老婆，而老婆所生的女兒，就是他們愛情的結晶,傳承了他最喜歡的女人的基因再混合遺傳了自己祖先留下來的的基因,這不是比前世的情人關係還要親密嗎？這要敲多少木魚,修行了多少千萬年才能得到的緣分啊！

　　雖然,後來變成我善良丈夫的好男朋友並不完全像我的父親，但是,在今生今世能夠做我父親的女兒已經夠幸運了，怎麼還能苛求要嫁到一個像父親的男人呢？

Visiting Friends By Boat.

Edited by Adela Karliner (12/27/2020)

One afternoon, Ann and Paul woke up from their nap. They were bored by sitting at home, so they called us and said they were thinking of visiting us by boat.

However, they were a little hesitant because they did not know how to navigate our house's water route.

Indeed, visiting a friend's house by boat and driving to visit a friend are two different things. What's more, we live on a half-salt, half-fresh brackish human-made river by the sea, bending deeply into the coastal marsh forest area.

We decided that we would take our small boat and go out to find their house. So, we hurriedly found the wooden oars, put on our caps, the sunglasses, safety vests, and the waterproof leather boots, and carried a map in a waterproof bag. The two of us hurriedly jumped onto the boat. By that time, the sun has begun to tilt westward.

Nearby, some fish jumped out of the water and soon merged back into the water again. The sound of our engine startled some waterbirds who flew out of the bushes on the shore.

Our boat moved from the human-made canal to the estuary, leaving water waves behind that crumpled the trees' reflections and houses in the water. We could see our neighbors' stilted houses and the coconut trees, pines, cypresses, and sawgrass in their yards move backward slowly.

When we got to the estuary, the intersection of saltwater and freshwater, we saw the water surface band with a subtle watermark pattern. When we passed through, my husband used our wooden oars to measure the water's depth, only about one foot! The voyage across the shallow water was very short. Three minutes later, we had entered the Homosassa River.

A new musician at the open-air bar had begun to tune on his guitar and adjust the microphone to welcome the evening guests on the riverbank.

When we gradually got into the freshwater, the shore's view was different from that brackish water. Each yard had an emerald carpet-like lawn. There are raging bougainvilleas, bright-colored hibiscus, and swaying green willows.

From their balcony, they had already seen us. Paul was waving to us. Ann ran downstairs, picked up the ropes of our boat, and tightened the lines on their dock pillars.

We disembarked, went upstairs, and into their house. We sat in the living room drinking Coca Cola and spread out the map. Luku pointed out that there was a shallow intersection at the river's sanctuary. It might be only about a foot deep, but their pontoon boat's air-cans were about a foot each. How could they pass there?

"By high tide, your pontoon might be able to make it." My husband has opened up a tidal rise-and-fall table to study.

They invited us to eat a light meal at their house, and we refused because we had to get back home before dark.

We had lived on Long Island, New York. There were lights all over the bay at night. In Homosassa, Florida, each household had a large yard. It is pitch dark at night; you could only see some light from the houses' windows, and the waterway was dark.

So they were afraid to ask us to stay too late, insisting that we take a lamp. We hurriedly jumped into our boat and headed home.

The open-air bar on the riverside had turned on all its lights. But in the afterglow of the sunset, they were still not very bright. So, when the boat turned into the canal where we lived, the tree's shadow started to appear to be a little fuzzy. When we got to the water in our backyard, the sky was not entirely dark yet. The sound of our small boat's motor had alarmed the neighbor's big dog. It started to bark. It also caused hidden, far away dogs to bark.

Back home, our home lights seemed very bright. After we bathed and found some leftovers in our refrigerator, we heated them in the microwave and ate them with relish. We both thought they tasted very delicious.

乘船訪友記

一天下午，安及保羅午睡醒來，在家閒坐無聊，打了一通電話給我們，說是想到我家來串門子。不過，保羅有點遲疑，他們由水面上不認我們家。

可不是，由水面上看岸邊的住戶，與開車從陸路去拜訪朋友家，好像是兩碼事。何況，我們住在海邊的半鹹半淡的人工河邊，彎彎曲曲深入沿海的沼澤森林區。

決定由我們先開了小船出去探一下路再說。連忙找到船上划的木槳，頭頂防晒帽、臉戴墨鏡、穿上充了氣的救生安全背心，足登防水皮靴，手上提了裹著防水袋的航船地圖，倆人匆匆跳上小船，這時候，太陽已經開始西斜。

遠處有幾尾魚跳出水面，很快又沒入水中，我們啟動引擎，發出突突的響聲，驚起岸邊一些水鳥，小船沿著彎曲的運河向河口開去，將水面劃出一股水浪，揉皺了映在水中房屋樹木的倒影，只見鄰居們的"吊腳樓"以及院子裡的椰樹、松柏及鋸草，一一緩緩向後退去。

到了河漠沙沙河河口，淡鹹水相交之處，只見一片寬廣的水面起著極細的水紋，老公將木槳伸進水中測量水的深淺，才一英呎左右！這段淺水航程非常短，三分鐘之後，我們已經進入河漠沙沙河水之中，河邊的露天酒吧裡，有個新樂手已經開始在試彈吉他，並調整著麥克風，準備接待入夜光臨的客人們。

河水漸漸變成淡水，岸邊的景色與鹹水岸邊不同了，每家院內有著像地毯一般的草地，有怒放的九重葛，有含羞的燈籠花，以及彎彎的垂柳。

由他們的陽台上，早就看見我們了，保羅一直對我們招手，安輕快地跑下樓來，接過我們小船的繩索，將繩圈在她家的柱子上。

我們下船，登樓進入沙家，坐在客廳裡一面喝可樂，一面攤開地圖，指點著告訴他們二條河鹹淡水交界處有一段路非常之淺，只有一英呎左右，而他們的汽筏上的汽筒也是一英呎左右，怎麼過得去。

"等到漲潮，汽筏就一定行得通了。"我先生又攤開一張潮汐漲退表來研究。

他們邀我們在他家吃便飯，我們不肯，因為天色已晚，怕趕不回去。

以前我們住在長島，長島灣內入夜之後，沿海都有燈火照明，佛州這邊地廣人稀，每家院子很大，天黑以後，只見樹欉中的房屋窗口透出一些微光而已，水面還是漆黑一片。

他們也就不敢相留，只堅持我們帶上一盞照明燈以防萬一。我們連忙又跳進小船，駛向回家的路途。

河邊露天酒吧的燈已開了，不過在落日的餘輝裡，還不太明亮，等小船轉進我們家住的那條運河的入口處時，樹影看起來已經有點模糊了，比及到了自家後院的水邊，天也還沒有完全黑，小船上突突的馬達聲響，已經驚動對岸鄰家的大狗，汪汪地叫了起來，極遠處似乎也隱隱傳來其他人家的狗吠對應之聲。

回到家中，只覺家中的電燈十分明亮，洗完澡，由冰箱中找了一些殘飯剩菜，用微波爐熱好，二人大口吃了，覺得分外可口。

The First Place Prize Winners

-Edited by Sandy Chin

One day, the phone rang loudly and I rushed to answer it.

"Gwen!" my mother cried on the other side of the phone, "I'm going to divorce your daddy!"

I was so shocked by what she said, I shouted at the top of my lungs, "Marriage is a lifetime event, Mom and Dad! Calm down, cool off, and think twice!"

"I've been thinking again and again." My mother was so agitated that her voice was even louder than mine. "I cannot cool down any more!"

"What's going on?" I asked.

"Since your daddy and I started studying English at the adult school, he has gone totally crazy. Every day we study from nine to seven with only a sandwich for lunch, and he still refuses to leave school even after the teachers have gone home. The night shift cleaner has to help us with our backpack every evening before starting his work. If this doesn't make you angry, what else does?"

The full name for the adult ESL class is "English as a Second Language" and it is a class to teach English to immigrants.

"Hey, mom, studying hard is a good thing. That's what you taught us since we were little!" I replied. Her current attitude toward studying English took me by surprise.

"But nobody is doing the cooking and cleaning at home. Dirty laundry is piled up as high as mountains...," my mother complained angrily.

"My dear father and mother, how about...?" I thought for a moment, and a great idea flashed through my mind. "From now on, we'll do what daddy likes to do in the morning which is to study English, and we'll do things mommy feels are important in the afternoon, like cleaning house and cooking. How about that suggestion?"

Mom's silence meant that she was in favor of my idea.

"This..." said my father, "this...," but he could not continue. So he also approved of the idea. I was very proud of my ability to mediate.

Unfortunately, before long, I received another phone call from my mother.

"Ah Ying, now, either I divorce your father or I will die," Mom screamed on the other side of the phone line.

"Slow down, mom, please slow down. What's up this time?" I asked.

"Your father is preparing for the English recitation contest for the adult class next month, so now he is forcing me to listen to his practice reading aloud over and over and over again until my spirit is about to crash!" my mother cried.

"Mom, if you don't like to listen, why not let others listen?" I tried to calm my mother down.

"Who are the others?" asked my mother.

"Neighbors, friends, classmates...," I listed a lot of possibilities.

"Well, if it continues, there will be no one who still wants to be our neighbors, friends and classmates!" Mom said with certainty.

"Things are really simple...," but I was cut short by my mother.

"Your fatal shortcoming is that your mind is too simple; reality is not as simple as you think," Mom yelled.

The reality really is not as simple as I thought because pretty soon, my sister the doctor also began to protest, saying our father could no longer practice reading to her, her nurse, or especially to her patients.

"My poor patients are already sick. How can they survive our father's recitations?" my sister asked.

Finally, we all agreed that our father should practice reading only to the tape recorder. This way, he could correct his mistakes by listening to himself and taking the recorded tape to the teacher for help.

My father's hard work eventually paid off. He outshone all of his competitors and won first prize in the English recitation contest!

To top it off, at the end of the year, my father actually graduated with the highest honors! Because my father's performance was so special, our mother received a special PHT (Push Husband Through) certificate to reward her great supportive efforts!

"Look at me, an old Chinese gentleman!" Dad said proudly.

"The credit belongs to the old Chinese lady!" smiled Mom very happily.

"Hurrah to the glorious Chinese people!" I said to my parents. We are so proud and happy to have a pair of award-winning parents!

阿公阿嬤得獎狀

一天，電話鈴驚天動地地响了起來，我火速飛奔過去接電話。

「阿英呀！」老媽在電話筒的那邊哭哭啼啼地說：「我要與妳老爸離婚。」

我吓了一大跳，急急喊道：「婚姻是終身大事，老爸老媽要冷靜、要冷靜，要三思而後行呀！」

「我已經思之再三。」電話裡媽媽激動得喊声比我還大：「冷靜忍下來了！」

「怎麼回事呢？」我問。

「自從妳替我們到成人學校註冊學習英文之後，妳老爸已經發痴啦，每天非要逼我也得朝九晚七，中午要一人一个三明治當午餐，晚上全校老師同學走光之後，他還不肯走，害得打掃教室的老黑清潔工每天都要幫忙我們整理好書包才能吸塵掃地，妳看氣人不氣人！」成人班的英語叫做 ESL 又叫 English as the second language 是給非英語的新舊移民學的。

「咦，用功讀書是好事，妳不是從小就這樣教導我們的嗎？」我詫異地問。

「家裡沒有人煮飯洗碗，髒衣服臭被單堆積成山....。」老媽怒道。

「老爸、老媽，你們看這樣好不好？」我想了一下，靈光突然一閃，說道：「以後每天上午做老爸要做的事，去讀英文，下年做老媽認為重要的事，整理家居、煮飯買菜，如何？」

媽媽不响，贊成了！「這個....。」老爸說。因為老爸"這個"不下去，也通過了。我對我調解的本領十分得意。

可惜得意不久，老媽又打來一通電話。「阿英呀，現在不離婚不行了。」老媽在電話裡尖聲地說。「慢來，慢來，又怎麼了？」我問。「老爸因為要准備他下個月成人班舉辦的朗誦比賽，成天逼我聆聽他朗讀他的英語文章，我的精神就快要崩潰了！」老媽在電話裡大哭起來。

「老媽，妳不愛聽，要老爸讀給別人聽，不就行了。」我息事寧人地說。「讀給誰聽？」老媽反問道。「鄰居、朋友、同學....。」我列舉了大量的可能性。

「哼，這樣下去，沒有人願意做我們的鄰居、朋友及同學了！」老媽很有把握地說。

「事情其實很簡單....。」我才張開口。「妳的缺點就是頭腦太單純，事實沒有妳想像的簡單。」老媽很生氣地罵我。

事情果然不簡單！因為後來我那做女医師的小妹也開始了抗議，說老爸不能對她、對她的護士、更不能對她的病人們朗誦。「可憐人家已經生病了，怎麼受得了老爸的朗誦呢？」妹妹說。

最後，足智多謀的我們一致公認老爸應該對著錄音機練習，不但自己可以聽了改正，還可以帶給老師改正。

皇天不負苦人心，老爸在眾多老中、老外的競爭中脫穎而出，得了首獎。更幸運的事，還在後頭，ESL 成人英語班裡學生有白、有黃、有黑也有西班牙，結業的時侯，老爸居然以優異成績畢業，因為老爸的表現特殊，老媽因此也得了一個 PHT 的獎狀，Push Husband Through，就是獎勵老媽幕後之功。

「看中國老先生多行！」老爸得意地說。「都是中國老太太的功勞！」老媽也笑嘻嘻地，非常高興。「咱們中國人太光榮啦！」我見兩位老人家像唱双簧一樣，就在一边附和凑趣，花花轎子人抬人嘛，不抬白不抬。

老爸老媽各得獎狀，真是人心振奮　，皆大歡喜。

My sister and my brother-in-law

Our Brother-In-Law John

In 1949, our landlord parents brought four hungry children, followed the government's hasty retreatment from mainland China to Taiwan. At the time, the local economy was very backward. There were more unemployed workers than available jobs. The life was tough. In 1952, my mother was pregnant again, and they were thinking of putting the newborn baby for adoption.

A month before the birth of our youngest sister, our mother changed her mind. She said: "It looks like the economic situation will get better! All we have to do is put an extra pair of chopsticks on the table; each one of us eat a little less food. We can keep our new baby!" Later, my father taught mathematics at Chia-Yi agricultural college, and my mother found a job as a librarian.

It was such a great decision made by our parents! Because our little sister was so sweet from day one. She often promise coquettishly to my mother: "When I grow taller, I will help my mom wash dishes. When I grow up, I will make money for my mother to spend. And I will marry a virtuous gentle son-in-law, together, we will practice filial piety to our parents." She is the close to heart little cotton jacket that warmed our parents.

Our little sister later turned into a stunning young woman, tall, thin with tender skin and thick hair. She was also very brilliant. She remembered anything she had read, was able to sing any song she had heard. Every time my parents looked at the certificates of merit covered our shabby wall and numerous prizes piled up our dilapidated bookshelf, they would say happily, " We are so lucky that we did not give her up for adoption. Otherwise, all these certificates will cover some other family's wall, and the prized will heaped up somebody else's bookshelf!" After graduation from Taiwan University Medical College, she became a famous pediatrician in San Jose and married a friendly and sweet dentist. They had two boys and a girl. My parents were so delighted and made many positive comments: "Are we fortunately that we did not give your little sister to others. Else, Ha ha ha!"

Probably out of the envy of the devil, our little sister had disappeared during a ski vacation!

From then on, our brother-in-law John had to work as a

father as well as a mother. He had to make a living, drive children to school, and help the kids apply for colleges. He was busier than ever.

Later, his oldest son Jason became a gastroenterologist, married a pediatrician Mabel, and the second Andrew and third Allison doctors of biochemistry and medicine. We congratulated our little brother-in-law John. He replied: "That's because of your little sister's genes are good!" We reminded him, "No matter how good the genes are, it should count as your credit that someone has to educate, tutor, and encourage them to became such achievers!" He heard it and felt it was true, so he smiled with joy.

John's three children are not only doing well academically, but they also have generous hearts. Once, my nephew, Jason, looked at a form I filled in. He said to me voluntarily: "Auntie Gwen, you may cross out your son   Descartes' name. I can be your contact person when something happens to you! Because he is too far away in San Francisco. Here is my name and phone number! You are closer to my office than his." His thoughtfulness and his kindness so touched me! In addition to inheriting his parents' good genes, of course, his father had brought up him the correct ways!

When John was busy taking care of his responsibility as a single parent, he had put off his personal affairs aside. It was not until his three children each had a successful career that he found a female companion to travel the world with us on the cruise ship. She was his childhood playmate in Hawaii. Ms. Chin is not only good-looking, in good health, and also has an amiable temper. We all love her very much.

Recently, during the prevention of the COVID-19 period, someone warned me and my other sister, Denise, "When your brother-in-law John gets remarried, he will not be your little brother-in-law any more!"

My sister Denise and I hear that we immediately answered in unison: "There will have no problem! As soon as he gets remarried, we'll immediately be extremely nice to his new wife and make her our bosom sister so that he will still be our little brother-in-law!"

My brother-in-law with his three children

My brother-in-law with his nephews and autor

我們的小妹夫

　　公元 一九四九年，我們的父母親帶了四個嗷嗷待哺的小孩，跟著政府倉促地撤退到台灣，那時當地的經濟十分貧乏，人浮於事，生活非常之困難。我的母親又懷孕了，打算孩子一生下來就過繼給人家。

　　一九五三年，在小妹生出來之前，母親反悔了，她說：「時局漸漸好起來了，家中飯桌上多放一副筷子，每人少吃一口飯菜，還是把孩子留下來罷！」

　　沒有想到這真是太好了，因為小妹不但長得漂亮，白嫩的皮膚，黑濃的頭髮，甜蜜的笑容，嬌滴滴的聲音，且特別聰明，書本讀過就記得，歌曲一聽就會唱，父母親每每看到家裡的錦標和獎品，就慶幸地說：「幸好當初沒有把小妹送給別人，不然的話，這些獎狀都掛到人家的牆上，獎品就存在人家的架子上了！」小妹台大醫學院畢業後，做了美國加州山河市的名醫，嫁了一位美國土生土長的牙醫，爸媽更是心滿意足地感慨：「妙手回春的女醫師是咱家的女兒，這麼溫和孝順的女婿是我們的東床耶！」

　　小妹從小就常嬌滴滴地對媽媽說：「等我長高，就幫媽媽洗碗，長大了，就賺錢給媽媽用，要嫁一個溫和賢慧的女婿，一同孝順爸媽。」她就是父母親貼身貼心的小棉襖。

　　小妹結婚以後，与小妹夫感情特別好，生了二男一女，一家五口，其樂融融。

　　大概是上天的忌妒，在一次滑雪渡假中，小妹妹竟然失蹤了！

　　從此，我們的小妹夫不得不父兼母職，上班賺錢，送孩子上學，陪孩子申請學校等等，整天忙個不停。

　　後來他的老大做了腸胃科醫生，娶了一位小兒科醫生，老二和老三都是拿到生物化學及醫學的雙料博士。我們恭喜小妹夫，那知他居然回答：「那是因為小妹妹的基因好的緣故！」我們正色提醒他：「不管基因多好，得有人教育、輔導、鼓勵他們才行，這些都是你的功勞喔！」他聽了以後覺得很有道理罷，就非常高興的露出一個孩子似的欣慰地微笑。

　　小妹夫的三個孩子們不但功課好，心地也非常善良，有一次我在填寫表格的時候，我的大外甥竟自動對我說：「大阿姨，你把表格上填的臨時聯繫人大表兄康康的名字刪掉罷，因為他在三藩市太遠，有什麼緊急事務打電話聯繫不上他，還是填上我的名字和電話號碼吧！我的辦公室比較近。」我聽了以後簡直感動到不行，世界上竟然有這麼細心善良的好孩子，除了繼承了小妹的好基因之外,當然是我們的小妹夫把他教育得特別好的原故囉！

小妹夫在忙碌中竟然把自己的終身大事給耽誤了。一直到三個孩子事業有成,他才找了一位女伴與我們一同去坐郵輪周遊世界，她是小妹夫小時候在夏威夷的玩伴，這位秦女士不但長得好看，身體健康，脾氣也非常和藹可親，十二分地令人喜愛。

最近新冠肺炎防疫期間,這位秦家妹子怕小妹夫一人寂寞，終於搬到小妹夫家去陪他了。

有人對我和大妹妹警告：「等你小妹夫再婚了，他就不是你們的小妹夫了！」

我和我大妹妹聽了，立刻異口同聲地回答：「這還不簡單，等他們一結婚，我們就立刻跟他的老婆結拜做乾姐妹，這樣，他還是我們的的小妹夫嘛！」

小妹和小妹夫

小妹夫和他的三個孩子

小妹夫和姪兒們

Celebrate the 61 Anniversary During COVOID-19 Pandemic

Edited by Adela Karliner (08/27/2020)

My daughter-in-law's parents had been married for 61 years. Their kids usually throw a big anniversary party for their parents every year.

Two or three weeks before the party, they would send out invitations, invite friends and family from all over the place to the city to celebrate.

Years ago, the paper invitation letters were sent to various locations by the post office. It was a paper card with beautiful designs and included the guests' names, the party's time, year-month-date and hour, and location. Sometimes, the invitation letter also had a stamped envelope and reply card; guests would answer the host whether or not they would attend, and how many of them would be at the party and so on. After receiving the return card, the host would know who and how many guests would attend to prepare in advance.

Later, the faster, environmentally friendlier e-mail took place. The electronic invitations could be self-made or purchased. They also let all invited people know who the guests were and included an electronic map, the restaurant's outside appearance, and detailed driving instructions. Such electronic invitation also was equipped for the guests to reply, so the statistic was public, it was not only convenient for the host to prepare, but it also let the guests know ahead of time, who would be at the party, etc., it is very suitable for everybody.

At the end of March this year, it was again the big day of their wedding anniversary. But, to prevent the spread of Covid19, we have all been quarantined at home. So not only did they use an e-mail invitation but only sent an electronic link.

After receiving the message, the guest can see the time between 4 and 5 p.m., the date of the party, plus the number of guests. There was no meeting place, let alone traffic instructions.

There were 15 people, all in front of their home computers, to participate in the zoom video. My in-laws have three children; each of them has two off-springs, a total of twelve people. My daughter-in-law is their youngest child.   She and her brother live nearby.

To prepare, the brother and sister arrived at their parents' home to turn on their parents' computers before the prescribed time.

Their elder son's wife was in her own home and shared a computer on a sofa with her younger daughter. Their eldest granddaughter was in Spain. Unable to fly home because of the worldwide pandemic, she could only join the party on-line.

Their elder daughter and her family live in New York. Today, her whole family sat at the table in a row. Here, I have to mention their elder daughter's son. He is a young elite who studies international relations at a famous university. At his young age, he has published his opinions in the New York Times. However, I have felt that his views about Sino-US trade were all from a white American perspective. The talented young man was going to attend graduate school at Tsinghua University in China. Then, the outbreak occurred; he could only sit home with his parents and sister to celebrate his grandparents' 61st wedding anniversary on video. I hope that he will go. I think it will open his mind to the views of some Asians.

My daughter-in-law rushed to her parents' house, leaving my son and grandson at home together. My granddaughter had just completed her college graduation trip to Asia in January, is consciously isolating herself in her apartment for two weeks. I live alone in Oakland, so I had to use a video.

The time had come; we all turned on our computer at home.   And, we, all 15 of us, could see everybody face-to-face happily. My daughter-in-law, the chairperson, called the meeting to order. She said some loving words to celebrate her parents' wedding anniversary. After her speech, we all took turns to express our sincere love for them and wish them many more years of celebration to come. The atmosphere was pleasant, joyful, and out of the warmth of our hearts.

After my turn to share my congratulations, I took out the cooked Taiwanese beef noodles from the microwave oven and picked up the noodles with my chopsticks very happily.

An hour later, we all shut down our computers.

Because of the video gathering, we were lucky enough to

take advantage of scientific progress. During the coronary virus pandemic, we could still be together and share our happiness. We were delighted to know that every one of us was healthy. Many years later, this will be a significant anniversary to remember.

Family Reunion

疫情中的結婚紀念日

　　我們的親家，也就是我媳婦的父母對於假慶節日非常重視，每逢他們結婚紀念日都是大開派對，在派對之前兩三個星期，就由郵局寄出邀請信，邀請了各地的親朋好友一齊到三藩市來大肆慶祝，可以算作一年一度的大喜慶。

　　多年前都是由郵局向各地寄出的紙質邀請信，紙製的卡片上有設計美麗的花樣之外，常常標明了；被邀請客人的姓名，聚會時間、地點。在那個餐館舉行，該餐館的名字，地址，等等。有時在邀請信裡面還慎重地附一封貼了郵票的回信信封以及回覆卡，客人事先回答主人來不來參加，有幾個人來參加等等，方便主人做準備。主人收到回卡之後，事先知道客人赴宴情況，以便預先準備。

　　後來，隨著科學進步，設計精美的邀請信改由伊妹兒 email 寄出，取其快速、環保且经济，電子卡片可以自製，也可以在網站上購買，填上被邀請客人的姓名，還可以讓所有被邀請的人知道，邀請了誰，以及人數的統計，也有聚會地點，另加電子地圖，餐館的外觀以及詳細的路途指示等等。這種邀請卡片，也有回信的設備，等客人回覆之後，就可以公開統計與会客人的統計數字，不但方便主人準備，且可以讓客人知道與他一同赴會歡聚的是些什麼人，大家都很方便。

　　今年三月底，又到了我們親家一年一度結婚紀念日的大日子，不過這一次是在防止冠狀肺炎蔓延的防疫隔離時期，雖然也是使用電子郵件邀請，不過信裡只有一個 link 連結 https://ucsf.zoom.us/J/792838970，收到的人點擊連結之後，就清楚的看到聚会是哪一天，幾點到幾點。一共有多少人參加。沒有聚餐地點，更沒有交通指示，也就是說，在指定的時間下午四點與五點之間，共一小時。人數呢，共有十五個人，都在自己家中的電腦面前參加視頻，我的親家有三個兒女，每人各有兩個孩子，共十二人。

　　小女兒就是我的媳婦與她大哥住倆老家附近。兄妹兩人在規定時間內，事先趕到父母家打開父母家的電腦，做準備工作。

　　大兒子的老婆在自己家，與她的小女兒同坐在沙發上合用一部電腦，他家的大女兒遠在西班牙，因為冠狀肺炎的緣故，不能坐飛機回家，也只能洲際視頻。

　　大女兒住纽约，全家四口坐在餐桌前一字排開，這提一下，親家大女兒的兒子，是一位學習國際貿易的美國名校青年才俊，常常在纽約時報裡發表文章，不過都是用美國白人的觀點來看中美貿易，這小子原定大學畢業後到中國清華大學進修，我是非常希望他去的，認為他到了中國後一定会打開他的胸襟，了解一些東方人的看法，可惜疫情發生，只能在家裡坐著與父母和妹妹，一同參加祝賀外公外婆結婚紀念日的的視頻。

我的媳婦趕到她父母家去了，剩下我的兒子和孫子兩人一同規頻，孫女兒剛剛完成到亞洲一月的大學畢業旅行，自覺地在公寓裡隔離兩週。我一人住在奧克蘭市，都只得利用視頻。

時間到了，在慶典中每人按照主席的提示發言，不外向兩位老人家說一些慶祝結婚紀念日的吉祥快樂的祝詞，在自己家中各隨己意想吃什麼就吃什麼，當著大家的面，由視頻中傳給大家看。輪到我說完祝賀的頌詞，逕自到廚房裡微波爐中取出煮好保溫的台灣牛肉麵來，一面用筷子將麵條叉到口中享用，一面笑嘻嘻地參加大家歡聲笑語。

一個鐘頭之後，大家一同關上電腦。

這次用各人自家的電腦軟體視頻歡聚，可以說是科學進步，在冠狀肺炎肆虐的時候，大家還可以因時制宜，歡聚一堂，人人看起來健康快樂，事後回味起來，也一定是一段佳話。

家族聚會

Free Lunch in Oakland, California

Edited by Adela Karliner (07/07/2020)

Free Lunch

My friend upstairs, Angela, sent me a message that the World Central Kitchen will be supplying lunch free of charge to anybody in Oakland, California, at 11 am every Monday, Tuesday, Thursday, and Friday between June 29th and July 31st.   The distribution place is at 310 8th Street.

Angela told me that to help those seniors who can't stand in line for a long time; she suggested that the food coupon be distributed from seven o'clock, early in the morning. After the citizens receive the ticket, they can bring the voucher to the distribution center to receive real food accordingly. I think this is a brilliant idea of caring for the seniors. Otherwise, when the weather is sweltering, and the sun is scorching, the long queues are indeed too much for the elderly, so I sincerely praised her.

Eighth Street? I live on 10th Street in Chinatown, Oakland, California. It is only two blocks away. I, an elderly lady, can come downstairs to take a walk. Because of the Covid-19 pandemic's current outbreak, I stayed home every day, cooking and eating with nobody but myself. It isn't exciting after a long while. I am so bored that I can feel that the bird is flying out of my mouth. Maybe it would be better to get a free

"

lunch cooked by somebody else and have a change of taste.

I usually get up and start to work at about five o'clock in the morning. On Monday, July 6th, wearing a mask, I went downstairs to find out the situation.

From Ninth Street, I saw two Asian women standing on 8th street outside the 310 building. They saw me walking over and immediately gave me a yellow lunch ticket with the number 11:10 am and number 49.

I came back home, got online, and checked. The World's Central Kitchen is a free kitchen initiated by a Latino, Jose Andre. They welcome volunteers and also donations. Although volunteer is the right choice, the elder's clumsiness may cause more problems than it's worth! I think that good deeds should be encouraged, so I filled in a gift within my ability.

At 11:10 am,   at the corner of 8th street, there set up a table with lunch boxes. After 11:10 am, I was the 49th person to receive two black plastic boxes filled with food.

Back at home, I opened the lunch box to take a look. There were three sections in the food container; one part contained the yellow cheese macaroni. The second section was a mixed vegetable of various colors: the cauliflower was white, the broccoli is green, and the carrot is orange. It not only looked fresh and beautiful, but it also tastes refreshing and delicious. The third section was the beef short ribs sauteed with onion and bell pepper, which smelt appetizing and tasted incredibly delicious. Although the short rib beef was a bit chewy, if you take time to chew it patiently, it would be fine.

After lunch, Ms. Marie, who lived upstairs in the next building, took a blank piece of paper and asked me to sign.

"Sign for what?" I asked.

"To complain that the beef short ribs are too tough! It is very hard for those who had bad teeth," was her answer.

"Come on! how can we complain about the food which was given to us for free? It is very ungratful!" I flatly refused to sign my name.

"Tell them what's wrong, meant to give them the opportunity for improvement. It is good for them," she explained.

After hearing this, I had to admit that she was right. So I

obediently signed the protest sheet.

I had decided to go early the next day, Tuesday, but unexpectedly got a light green 11:15 am No. 15 coupon ticket; you could interpret it as that everybody had the same idea; everyone liked the meals; many people had decided to arrive at the place to get the lunch ticket earlier. So I decided to leave the free lunch for those who need it more than I do and stay home to cook for myself.

On Wednesday, September 16th, about 11:30 am, my doorbell rang. During the pandemic prevention, I stayed at home most of the time and was bored to death, so I immediately rushed to open the door.

"Hello, this September and October, we are distributing free lunch to the homebounds every Monday, Wednesday, and Friday, do you want one?" asked my next-door nice-hearted neighbor who was pushing a food cart with a pleasant smile.

Ah, the seniors could get free lunch without standing in a queue, what a great idea!

P.S. . The World's food Programme has won the Nobel peace prize for the year of 2020, congratulation! WFP well deserves the honor and the money!

加州奧克蘭市的免費午餐

免費午餐

　　樓上好朋友安琪拉傳來消息說：世界中央廚房將在六月廿九號和七月卅一號之間，每週一、二及四、五上午 11 點，在加州奧克蘭市免費供應午餐，發放地點是 8 街 310 號。

　　據安琪拉說，為了避免一些老先生和老太太不能久站排隊起見，所以她建議由七點鐘就開始發送領取食物的票卷，市民拿了票卷以後，就可以按照上面的號數，正式領取實質食物。我覺得這真是一個貼心愛民的聰明主意，不然天氣這麼热，太陽這麼大，排起長隊來，的確是令年紀大的長者吃不消的，所以由衷地稱讚了她。

　　第八街？我住在奧克蘭中國的中國城的第 10 街，只有兩街之遙，老太太下樓散一下步就到了。自己尋思，目前因為冠狀肺炎的疫情，每天窩在家裡中，自煮自吃，日子久了，無聊到極點，口中淡出鳥來，不如去領一份免費午餐回來換換口味。

　　我平常清晨五點鐘左右就起來工作，這個星期一，七月六日，一大早到了七點鐘，就戴了口罩，下樓去一探虛實。

　　走到第九街，果然就看見兩個東方女人站在三佰一拾號外面的街頭，她們見我走過去，立刻發了一張上面寫著 11:10am 49 號黃色的午餐票給我。

　　回來上網查了一下，世界中央厨房是由少數民族拉丁裔的荷西·安德烈發起的免費厨房，他們不但欢迎義工，同時也欢迎捐款。做義工雖是上選，但是老人家行動不利索，萬一打翻食品，弄巧反拙，

若是跌了一交，就更加得不償失了！總之，善行應該鼓勵，就按
自己經濟能力，填了一個能力之內的捐款。

中午十一點十分，第八街三佰一拾號的門前街上，果然支起
了一個發放午餐盒的桌子，上面擺滿了午餐，我是第 49 個領取
午餐的人，領到了兩個黑色裝了食物的塑料盒子的午餐。

回到家中，打開午餐盒來看，原來盒中裝了一格黃色起士通
心粉，第二格是各種顏色混合的蔬素菜，白的是花椰菜，綠的是
綠花菜，黃的是胡蘿蔔，混合得不但看起來新鮮美觀，吃起來也
非常清爽可口。另外一格裝的是洋蔥甜椒炒牛仔骨，真是鮮美可
口，雖然牛仔骨肉是有點硬，多咀嚼幾下，還是可以下咽的。

吃完免費午餐，隔壁樓上的瑪麗小姐拿了一張白紙叫我簽名，
指責牛仔骨太硬，牙口不好的人吃不動。我起先不但不肯簽，還
辯說：「人家是免費的，怎麼可以挑精選肥呢？」她當時就正色告
訴我：「告訴他們缺點，他們才有機会改進，是對他們好的。」
我聽了以後覺得有點道理，就乖乖地簽了名抗議。

決定第二天,星期二早一點去，不料反而得到了淡綠色
11:15am, 第 15 號的票子，可見人同此心，大家都覺得餐食內容
不錯,很多人都決定早一點抵達分發午餐票的現場了，人數一多，
排的隊伍自然就比較長了。當然，也可以解釋做有很多人需要這
種免費午餐，所以,我決定放棄去排隊領取午餐。

九月十六日星期三，大概在上午 11 點半左右，我家門鈴響
了，防疫期間，天天在家中悶坐的我立刻飛奔过去開門。

「你好，奧克蘭今年九月、十月,每逢星期一、三、五日，派
送免費午餐到各家，你要一份嗎？」原來是我隔壁有善心的鄰居
穿了義工的背心，正在笑容滿面地推着滿車的食物向各家各戶派
送免費午餐。

啊，原來好人善事做到底，老人家們連排隊領餐都免了，真
是好主意！

後記：本年十月，世界糧食計劃榮獲公元２０２０年諾貝尔
和平獎，實至榮歸，人同此心，心同此理，受之無愧！！

Mobile Home Of Love

Edited by Adel Karliner (12/27/2020)

In general, all Americans have an instinct for beauty. Carpet-like green lawns and colorful flowers surround homes in both exclusive communities and houses in rural areas. Even prefabricated houses have neat and beautiful yards. That includes the cheapest mobile homes with no exception.

Whenever I drive past the mobile home park, I see this lovely mobile home without glancing at it. The house owners must have a fairytale-like vision; this royal blue coarsely painted old moveable house has white windows and door frames. A row of bright yellow sunflowers stands happily on the north side. They are all taller than a human. Early in the morning, more than a dozen of these golden flowers face the rising sun. They all turned to the west in the evenings. Not even a single one of the flowers is in the wrong direction. It is absolutely a small house in the wonderland.

The mobile home is tiny. The door frame has a simple plastic rain shelter. Outside the door, there is a wooden colored balcony connected with a staircase. The small house leads to a wasteland-reclamation garden with vibrant, colorful flowers. Outside the park, there are a dozen toylike pink plastic fire crane flamingos. The gravel-covered the land next to the house; they put tree branches in the ground and then covered the four pillars with a plastic shed. It would probably look quaint. They have used the long leaves of cut coconut trees on the top of the plastic tent.

The owners are hospitable. At noon, when Florida's sun is high and hot, outside the house under the shed, there always are people taking naps. In the cool evenings, the yellow lights of the electric bulbs repel mosquitos. The moon shines, the dark sky sparkles with twinkling stars at night—the small Christmas light bulbs on the grass shed's roof flash in the cold winter. Next to the outside bonfire, there always has several friends drinking and talking loudly. The place is full of warmth and laughter.

Once, passing there, I saw an old truck parked near the house. Somebody had just pushed open the truck door. The woman who was watering the flowers turned off the faucet in her hand suddenly. She dropped the hose on the ground and flew to kiss a soiled worker who came out of the truck's open door.

Wow, the owners of the house are so young! I thought.

One day, my doorbell rang, and I ran over to open the door.

"Mrs. Li, here's a letter for you!" In the countryside, the average house has a huge yard. The driveway is very long, so the mailbox is usually located on a public road. The postman or postwoman stops at the mailbox, stays in his or her car, opens the car window drops the letter into the inbox.

"Thank you for bringing the letter in, are you the postwoman? Why aren't you in uniform? How come you are not driving a postal car?" I asked. The woman who delivered the letter looked familiar.

"I'm a substitute postwoman. We get the job only when the post office is too busy. So I don't have to wear a uniform, and I don't get a postcar to drive," she replied.

We have two post offices in the town of Homosassa, Florida, a new and an old. The old one is near my home, and It only has one regular staff member. And it only receives letters. They don't do delivery. So, the woman must be a substitute for the new post office.

"I pass your mobile home every day. You have dressed up your place beautifully! It must take a lot of work to keep it the way it is, right? How did you think of being a substitute postwoman?" I suddenly remembered that she was the fairytale mobile home's young mistress and was very happy to ask her.

"Thank you for your compliments! we also like our house, it is perfect for the two of us. He does odd jobs, twenty dollars per hour, it also is enough for two. But we soon will have children, and with children, that mobile house will be too small, so it is necessary for us to save some money. As a substitue postwoman, I make thirteen dollars an hour. Of course, I don't often have a job, but it is not without a small supplement. When we have some small additions, we're going to sell this tiny house and buy a bigger place," she told me.

Really! This small love nest not only has flowers, grass, stars, moon, warmth, laughter, and love, there is a more glorious future hope too!

愛之小車屋

　　一般的美國白人愛美成了天性，不要說一些高級社區的庭院一概綠草如茵，繁花似錦，連在南方窮鄉僻壤的地方，由工廠預先造好一幢幢的成屋 prefabricated house 也經常是房屋整齊美觀，房外花木扶疏，甚至最廉價的行動車屋 mobile home 也不例外。

　　每次開車經過那裡，一眼就看見這幢可愛的小車屋。主人一定很有童心，這幢舊車屋被粗粗地漆成寶藍色，窗框及門框卻是白色的，窗外北面種了一排比人還高的向日葵，清晨日出，十幾株金黃色大朵的花兒全部朝東迎著初起的朝陽，傍晚又轉向西方，沒有一朵花兒開錯方向，簡直就是童話中的小小房屋。

　　車屋其實很小，門框上有一個簡單的塑膠遮雨蓬，門外搭了原色的木製陽台，陽台連著一條原色的木樓梯，通往一塊荒地上開墾的花園，裡面熱熱鬧鬧地開滿了千紅萬紫，園外圍了十幾隻好像玩具一般粉紅色的塑膠火鶴 flamingo，旁邊是一塊碎石地，地上用樹枝搭架，架上撐著塑膠棚，大概為了要顯得古拙原始，所以塑膠上還覆了不少由椰子樹剪下的長長的葉子。

　　主人好客，大太陽天的樹蔭下，棚裡棚外都有人在憩睡休息，傍晚避蚊火照著乘涼、談天、喝啤酒的三、四位朋友，天黑了，有星星也有月亮，冬天房屋屋沿的小燈泡一直連輳到草棚的棚頂，棚邊熊熊地燒著營火，有溫暖也有歡笑。

　　有一次經過那裡，正好看見一輛舊的小卡車停在屋邊，正在澆花的女人突然關掉手中水管的水源，將水管丟在地上，飛奔過去擁吻由車門開處走出來的一位工人，原來男女主人這麼年青。

　　我家門鈴響，我奔過去開門。

　　“李太太，這裡有一封妳的快信。”申明一下，我們鄉下，一般人家院子極大，車

　　道極長，所以信箱就設在路邊，郵差開了郵車，坐在車上將平信投入信箱就行，不必繞進車道，車道極長，太花時間。

　　“哦，謝謝妳將信特地送進來，妳是郵差嗎？怎麼沒有穿制服？也沒有開公家的郵車？”我問，這位送信的女子非常面熟。

　　“我是後補郵差，只有郵局太忙時，才由我們後補送信，所以不必穿制服，也沒有

　　郵車可開。”她答。我們佛羅里達州的河漠沙沙鎮上有一新一舊兩個郵局。舊的那個在我家附近，裡面只有一位工作人員，每天中午他午休時郵局就關門一小時，他休假時就由另一位女工作人員代理，舊郵局只管收信，不管送信，這位女子一定是新開的那家郵

局的後補郵差。

　　“開車常常走過妳的車屋，打扮得好漂亮喔！一定花不少工夫吧？怎麼想到要做後補郵差來送信呢？”我突然想起來她就是童話中車屋的年青女主人，非常高興地問她。

　　“謝謝妳讚美，我們也很喜歡我們的房屋，兩人住正好，他打零工每小時廿元 兩人也夠用了。可是我們不久一定會有孩子，有了孩子的話，那車屋就太小了，所以必需存一些錢，後補郵差每小時約十三元，當然不是經常有工作的，可不無小補哦。我們打算有了小寶貝就要將車屋賣去，再添些錢買一幢大一些的住處。”她告訴我。

　　真是，原來那幢小車屋，不但有花有草，有情有愛，有太陽星星月亮之外，還有更大的憧憬和希望呢！

花園的一角

Hitting The Jackpot

Edited by Cathy Chapman (2020/04/05)

Dad and Mom went to do their grocery shopping. When they pushed the shopping cart to the cashers and was about to pay for their groceries before they get out, suddenly, in the supermarket, the alarm went off, the lights flashed, and there were a lot of small pieces of colorful papers started floating around, like snowflakes fell on my father's and mother's face, head, clothes even on the shopping cart.

The supermarket manager came out, shook hands with Mom and Dad vigorously, and announced loudly that these two elderly seniors were the 100,000th customers of their market and that there is a special deal today. All the customers in the supermarket crowded over to watch the excitement.

"The two of you, please unload your grocery in your cart and start the shopping all over again. All the food that can be stowed into your cart and reached the cash register within three minutes will be 'on the supermarket,' that is, will be free for you to take home!" The manager raised his hands and made a loud and clear announcement.

The crowd's spirit inside the supermarket was boiling, the applause was thunderous, and the cheers were loud enough to reach the sky.

The purchasing starts! Dad is inclined to grab the sausage and ham. However, mom thought the dried shrimp and mushrooms are more delicate. The tacit for mutual understanding between them was not very good. Two of them pushed one cart, each pushed toward different directions, since father was stronger, the carriage went to the counter of ham and sausage, after dad took the processed meat, he also grabbed some dried Abalone and sea cucumber. The shopping cart pulled my mother to a counter where she held two bags, inside each bag, there were various kinds of gorgeous packaging.

Come on! Come on! Their cheerleader's applause is amazingly loud.

Ham, sausages, mushrooms, abalone, were all very welcomed by their old friends in the senior citizen center, but no one had asked for any of the treasures snapped up by mother.

What did she pick? So gorgeous and exquisitely packed?

"Mom, how are you going to use so much cheese?" I asked. My husband was the one least fond of eating cheese. He called cheese in Chinese is 'die from angry.'

Suddenly a light bulb switched on. Smartness suddenly hit me! My younger brother-in-law was growing up in the United States, and he might be able to make these two huge bags of treasures not to be wasted.

I started to feel very happy.

I drove my car to my sister's house with the cheese.

When I rang the doorbell, my sister's mother-in-law Po-Po came to answer the door. Po-Po carried her newborn granddaughter on her back.

When she saw that her daughter-in-law's elder sister come to her house to visit, she was pleased. And then saw these two bags of her son's favorite expensive cheese, so she gave me an extra welcome smile. She and her HongKong maid came back and forth in the kitchen and steamed many HongKong-style snacks for me to eat. She also boiled a pot of freshly made hot chrysanthemum tea for me to drink.

We spoke different Chinese dialects. Our words were not very clearly understood. There is a communication problem between us, so we were all heaped with smiles to show kindness. After smiling a long time, my face felt so sore as if I have chewed three and a half pounds of beef jerky.

Po-Po also loaded two big bags for me to take home to show her courtesy. She opened one of the bags, picked up a piece of the food with chopsticks, and personally sent it to my mouth, "Auntie, please taste my yu-Je (oil chicken)." Her Chinese had a strong Cantonese accent.

I took a bite. It was a piece of barbecue pig!

As I tasted the pig, which was called the oil chicken in Cantonese, I pushed the lid on the other bag with a large black jar inside. "What is this? What's it for?" I asked politely, of course, my spoken Chinese was with heavy Taiwanese accent.

"Tonics, it is the tonic for your body." She answered in Cantonese.

I thought since the Cantonese called barbecue pig "oil chicken," so there was no need for me to investigate the name of the tonic supplement, of course, and, there was also no reason to do any further investigation, any new research would yield nothing anyway.

Now, the result of friendly exchange was that I gave away two bags of beautifully packaged cheese, received a sizeable black jar of tonic which I could not name, plus a large container of food which its name did not match the actual stuff.

Sure enough, there were always smarter people among the brightest, when I left my sister's house for home, Po-Po personally walked me to my car, and asked: "Next time, if you win more cheese, please send it over again."

"From now on, I'll go to this very supermarket to shop too. Maybe I also have the same luck to grab a lot of free cheese for my son!" Po-Po also made up her mind.

阿公阿嬤中彩券

　　老爸老媽買好菜，將手推車推到菜場門口，正要付錢出門時，菜場內突然鈴聲大作，燈光閃爍，門的上方飄下大量的五顏六色細碎的彩紙，像雪花一般的落在老爸和老媽的臉上、頭上、衣服上以購貨的推車上。

　　菜場的經理與老爸老媽用力地握手，大聲地宣佈，說兩位老人家是該市場的第十萬位顧客，獲贈彩卷一張，所有的顧客全都擠過來看熱鬧。

　　「請兩位老人家將車再推回去，重新搶購。凡是能在三分鐘之內裝入車內推到收銀處的食品，憑此彩卷，一概由敝超市免費奉送。」這位經理舉起雙手慎重的宣布。

　　超級市場內群情沸然，掌聲雷動，歡聲震天。

　　搶購開始！老爸一心要拿香腸火腿，老媽卻覺得蝦米香菇比較爽口，兩人默契不夠，一輛購貨車兩人同推，各推各的方向，結果老爸力氣大，車子只好向火腿香腸的方向走，爸爸忙著拿火腿香腸，又順手取了一些鮑魚海參，老媽剛好被車子拖到另外一個櫃前，只得就地取材，抓了不少華麗包裝的大包小袋。

　　加油！加油！老爸老媽的啦啦隊陣容驚人地浩大。

　　火腿、香腸、草菇、鮑魚，他們在老人中心的的老朋友們都很歡迎，只有媽媽搶購的珍品，無人問津。

　　她老人家抓了些什麼？包裝得如此華麗精緻？

　　「老媽，妳怎麼拿這麼多起士乳酪呢？」我問。我的先生最不愛吃起士乳酪，硬是把"起士"叫做"氣死"是絕口不嚐的。

　　還好，我靈機一動，想起小妹夫是在美國長大的廣東華人可以使這兩大包珍品不致浪費，心裏很是高興。

　　開了車，帶了兩大袋起士乳酪，到小妹家按門鈴，替小妹管家的婆婆揹了在美國出生的孫女兒前來應門，見是媳婦的姐姐登門造訪，十分高興，再看見這兩大包她兒子最愛吃的昂貴起士乳酪，更是眉開眼笑，她老人家與她家的香港女佣在廚房中來來回回，蒸了不少廣式點心給我吃，又泡了滾燙的廣式菊花茶給我喝。因為言語不通，溝通不夠流暢，所以我們都堆滿了笑容以示親善，笑了一陣子之後，臉部就好像嚼了三斤半牛肉乾一般地酸痛了起來。

　　妹妹的婆婆也裝了兩大包東西要我帶回家，以示禮尚往來，她老人家打開其中一包，用筷子夾了一塊，親自送到我口中。

　　「請阿姨嚐嚐我自己燒烤的油雞。」她老人家的國語有很濃重的廣東腔。

我咬了一口，明明是燒烤乳豬。

我一面品嚐那叫油雞的乳豬，一面用手推推另外一包裏面的罐子上的蓋子，裏面烏烏黑黑一大罐。

「這是什麼？做什麼用的呢？」我很禮貌地笑著問，說的當然是台灣式國語。

「補品，是進補用的。」她老人家用廣東式的國語回答。

我想既然燒烤的“乳豬”廣東人硬要叫他“油雞”，至於補品的名稱，也不必深究了，對我們這批說台灣國語的人來說，深究了不也是白搭嗎？

現在，投桃報李的結果，是我送出了一大批花花綠綠高級名牌包中的“氣死”，帶回去一大罐烏烏黑黑說不出名稱的補品，以及一大包名不符其實的燒烤。

果然，強中自有強中手，回家的時候，婆婆親自送我上車，還不忘很客氣地用夾著英語的廣東國語吩咐我：「下次親家們再有彩券搶購起士乳酪的話，請再送過來。」

「從今以後，我也要到這家庭超級市場去採購，說不定我也中一張幸運彩卷哩！」她老人家也下了決心。

A Lively Riverboat Parade.

Edited by Adela Karliner (12/29/2020)

On the Saturday before Christmas, all the residents in the town of Homosassa were in high spirits. On the river, at night, there was a lively parade of boats with beautiful lights.

At about five o'clock, my husband and I hurriedly drove to the Wangs' house to eat Wang's homemade Taiwanese style braised beef noodles. After eating the warm soupy noodles, the four of us went together to the Cohens. It was getting late. All the colorful Christmas lights that hung along the eaves of residents' houses had begun to shine.

The six of us got into the Cohens' big station wagon and headed to the Sergeants'. From a distance, we saw the Sergeants' house had all their lights on. There was a long row of cars parked on the roadside. It was not easy to find an empty place to park our big station wagon. Climbing up the Sergeants' stairs, we saw that many small light bulbs wrapped up their stairs' rail. It was very cheerful. We pushed open the door and could see the colorful Christmas lights that decorated the hall. The room was full of people talking, laughing, and merry-making. Lotus brought out six cups of hot tea for us.

As everyone sat down and was about to start drinking tea and snacks, someone outside shouted, "Come on, come here, the parade's light fleet has begun to move!"

The people in the room were all in high spirits; hearts began to beat faster. Immediately the guests filed downstairs.

Lined up on the Sergeants' deck, we saw a brightly lighted ship moving slowly from upstream to the open river estuary. Next to many colorful lights decorating a giant Christmas tree on the leading boat stood three saints. And the ship was followed by a vessel with colored lights tied onto three fawns towing a beautiful multicolored carriage, on which sat the Santa Claus with a microphone shouting: "Ho, Ho, Ho!" The unique one was a large truck decorated with lights on a large boat. The vehicle even had four glowing wheels rotating. Following this boat was a smaller truck also with brightly rotating wheels on a smaller boat. Both vehicles' megaphones played loud American Christmas songs.

Last but not least came a row of three small ships with all kinds of bright decorations. Although we could not decipher the detailed stories they were trying to convey, the degree of happiness was no less than other gorgeously lighted boats.

In ancient times, there were flower boats on the Qinhuai River in China at night. What was the event? To what extent was the light shining? Of course, those boats' speed must have been slower because now we use the motors to move the Homosassa Riverboats.

"The local newspaper reported that there were twenty-eight ships signed up for this year's fleet parade competition, and look! each one is passing the judges' deck now!" Frank was pointing at the boats.

"One, two, three... twenty eight, not bad, a total of twenty eight, have passed by, and each has its own merits. " Each of their designs was ingenious.

"Quick, quick, we need to get to our neighbor on the left side, the Pompeos' house, before the light fleet turns around," Lotus lowered her voice and shouted.

"Why? Isn't the light fleet coming back on the original route? "

"The Pompeos have specifically bult a fire on their riverbank for us. We should do our best to live up to their good intentions! So, we must get over there before the light boats turn around."

When we arrived at the Pompeos' house, the bonfire in front of their riverbank was booming. The campfire drove away the cold of the December night. The Pompeos saw us coming, quickly moved chairs, and poured drinks with welcoming smiles.

"Next year, we Chinese will also sign up for the light boat competition, we can use lanterns to decorate a dragon boat, and we can certainly win," Wang said suddenly.

"Yes, yes, we can have the dragon spit fire from its mouth." There was a close second.

"The spitfire dragon is a Western thing, we Chinese should have two dragons fighting for one ball and some other Chinese quitessence of Chinese culture!"

After watching the last of the twenty-eight light boats turn around and head upstream, we went noisily back to the

Sergeants' to eat snacks, drink hot tea, and sing karaoke. We were still talking, singing, laughing, in full swing when it came time to go home.

On the roadside of Highway 19, several flashing police cars stopped the passing vehicles. They were there to measure the drivers' alcohol consumption. Traffic became extremely slow. The roadside police car flashed alarming lights, as they checked the degree of the drivers' drunkness. It is part of the United States holiday landscape.

The riverbank in our backyard leads to the Homosassa River, but enjoying the music alone is not as fun as enjoying the same piece with your friends. The two of us watching twenty-eight boats passing by our backyard certainly was not as much fun as watching along with our neighbors. Human beings take this opportunity to gather together, burning campfire to drive away the winter cold. Eating, drinking, talking, singing, and laughing with friends because human beings are social animals.

熱鬧的河上燈船大遊行

　　聖誕節前的星期六，河漠沙沙鎮的居民個個興奮異常，因為入夜的時辰，河漠沙沙河上有一個熱鬧的燈船大遊行。

　　五點鐘左右，老公與我就急急忙忙開車到王老闆家，吃王老闆親自燒煮的臺式紅燒牛肉麵，大家吃得暖乎乎，笑嘻嘻，然後一同開車到柯家，此時天色已漸晚，一路上每家每戶掛滿的紅綠彩燈，已經開始發亮。

　　我們與柯家夫婦，一共六人坐進柯家的大旅行車中，老遠就看見沙家的樓房上張燈結彩，在路邊排著的汽車中，好不容易找到小塊空地方將大旅行車停好，向沙家的樓上爬，只見樓梯的扶手上都纏掛著各色小燈泡，真是琳瑯滿目，推開大門，裡面被聖誕節的彩燈裝飾得五光十色，美麗非凡，房內已經有很多人在談天說笑，一進門，蓮就連忙泡了兩杯熱茶給我們。

　　大家寒暄完畢，坐下來正要開始喝茶吃點心的時候，室外有人大喊：「過來了，過來了，遊行的燈船隊已經開始出動了。」

　　室內的人們精神一振，心跳加速，立刻魚貫下樓，在沙家水邊的甲板上站穩，黑暗中，一艘極大的船已經由上游慢慢地向河口開了過來，船上用彩燈紮成一棵極大的聖誕樹，旁邊有很多小燈，大概是三位聖人之類，這一艘才過去，又來了一艘船，上面用彩燈紮成三隻小鹿，拖著一輛馬車，上坐聖誕老公公，用擴音器對著河上大喊：「呵，呵，呵！」，其中最別出心裁的是一艘大船上用燈裝飾成一輛大卡車，卡車甚至有四個發光的輪子在轉動，後面還拖了一輛小船裝飾成的小貨車，車上擴音器播放著美國民歌，再過來一排三艘小船，上面各式燈飾，雖然看不懂是什麼，但其熱鬧快樂的程度，比其他明亮的燈船毫無遜色。

　　古時候，中國秦淮河上的花船，入夜以後，不知什麼盛況？光明燦爛到什麼程度？當然，速度肯定比較慢，因為現在的船上都是用機器馬達來推動的。

　　「鎮上的地方報紙登了消息，說今年河漠沙沙報名參加公開燈船比賽的遊行船隊一共有廿八艘船，看，他們一一駛過前面裁判的評審台。」法蘭克指指點點地說。

　　「一、二、三．．．廿八，一點不差，一共廿八艘，每一艘都走過了，且都各有千秋。」駛過去的船隻，艘艘爭奇鬥艷，每一隻都各具匠心。

　　「快，快，我們要在燈船隊調轉頭來之前，趕到左鄰柏家。」蓮壓低了聲音喊道。「為什麼？燈船隊不是要從原路回來嗎？」

　　「柏家為了我們，特地在河邊燒了營火，不能辜負人家的好意，所以一定得過去那邊看繞回來的燈船。”

　　到了柏家，他們燒的營火正旺，趕走了十二月夜間的寒氣，柏家見我們到來，笑逐顏開，趕忙搬椅子、倒飲料，非常熱烈地招待我們。

　　“明年，我們這裡的中國人也去報名參加燈船競賽，我們用彩燈紮出一條龍船，一定可以旗開得勝。”王老闆突然說。

　　“對，對，可以口中噴火。”立刻有人附議。

　　“噴火龍是外國玩意兒，我們中國人要來個雙龍搶珠之類的中國國粹！”

　　在柏家看完廿八艘燈船全部繞回上游，大家又熱熱鬧鬧地回到沙家去吃點心及熱茶，又同唱卡拉 O.K.，一直談天唱歌，笑到很晚，這才興猶未盡地各自散去。

　　到了十九號公路邊，好幾輛閃著警燈的警車，攔住來往車輛，測查駕駛人口中的酒氣，以及酒醉程度，使得交通變得極慢，我想，這種公路邊警車閃著警燈，查測汽車駕駛員的酒醉程度，也可以算著美國假期節日的一大景觀罷。

　　我們家後院的小河，其實是通到大河中的，但是獨樂樂不若與眾樂樂。若只是自己倆人孤獨地在河邊觀看廿八艘燈船在河上先後行駛，一定沒有什麼意思，非得有一大批朋友，眾人借這個機會聚在一齊，燃燒營火驅寒，一同吃喝說笑，評頭論足，這才熱鬧有趣，因為人類是合群的動物啊。

The Handsome Guy Lives Across The Hall

Edited by Sandy Chin

When Ruth opened her apartment door the   first morning after moving into Oakland Chinatown's new home, she was taken by a big surprise! Guess what? Because there was a handsome guy stood right in front of her! although you cannot tell the original color of the hair on the top of his head, but you can be very sure that the eyes behind his glasses are particularly handsome.

"Who are you?" Asked Ruth very politely.

"I am your neighbor lives across the hall." The good looking guy answered.

"Can I be of any help?" She felt that she is so lucky to have a handsome guy as her neighbor, and he is actually standing in front of her apartment.

"I am locked out my own house!" Wow, what a nice voice he has.

"What can I do to help you?" again asked Ruth.

"Thank you for your concern, I have called a locksmith, they are on their way."

"That ···, do you want to sit down and wait?" Ruth quickly ran home, took a chair out for him to sit.

"Thanks, That's very nice of you!" he said, and sat down outside his own apartment, i.e.,in front of her door.

"Do you want my phone number?" Before returning to her own apartment Ruth asked again, then, she felt something was not right, so she quickly added: "So, if you want to call me you will be able to call me."

What she just asked really made herself very uneasy; You don't ask a stranger does he want your number! It is so inappropriate, what if he misunderstood? However, what has been said had been said, there was nothing you can do about it anymore.

Two days later, in the elevator, she met the handsome guy again, ah; the opportunity to clarify is finally here. "I'm sorry, last time I asked you if I you want my phone number, because it was ..." Somehow, it seems to her that she has made the bad situation to worse.

"This is my phone number, and if you need me, please call me!" He did not wait for Ruth finishing her explanation; immediately took out a business card from his pocket, and gives it to her.

Ruth looked at his business card, it not only has his phone number,415-391-7767

but also has his office address, as well as email address, even his website, www.DownToEarthlawyer.com

Ah, and he is a lawyer! She now is very happy, immediately called a new friend of hers; "Beautiful Flower, there is a lawyer across the hall from my apartment, may be you can ask him to handle your divorce!"

What has really happened? Ruth's new friend Beautiful Flower whom she met in the elevator, told Ruth about her own story that her less than one year new husband who is 12 years younger than her is keeping his cell phone 24 hours in his hands, and he is not only pick up his ex-wife's son from school every day, but also secretly seeing his ex, so Beautiful Flower has found a private detective to investigate his daily activities behind his back, all the pictures has been collected as the evidence.

"He has never loved me, it is clear that he want to use my marriage to him in order to stay in the United States." Beautiful Flower cried to her new friend Ruth: "Now, what else can we do but to get a divorce?"

However, Beautiful Flower and Ruth has never going through a divorce, and do not know how to start a divorce procedures, and now, since there is a lawyer lived across the hall, of course, they are not going to miss this good opportunity!

So, Beautiful Flower use Ruth's Kitchen to cook a great Chinese meal and Ruth went across the hall to ring the handsome lawyer's doorbell and invited him to have a Chinese lunch with them, and ask him to handle Beautiful Flowers' divorce case.

"I'm sorry! I'm not a divorce lawyer!" Said the guy. Ruth and Beautiful Flower looked at each other speechlessly, they did not know the division of the of the lawyer's job is so fine!

"I will try first; if it doesn't work then I will introduce Ms.

Beautiful Flower to my friend who is a divorce lawyer." He thought for a moment, and then said.

It turned out that this good looking guy is not only handsome but also has a good personality; gentle, polite, and friendly, with many male and female friends, where "s" means plural, in addition to the divorce lawyer, They met his young black tenant and many other friends, such as Chinese sculptor, American literati, black artists, … his home is like a international melting pot, very Warm and happy.

Because Ruth does not drive, so he took Ruth to the grocery shop to buy food, they went together to the city hall to fill out forms, sometimes went to the restaurant to enjoy good food, and chat happily together.

There is a Chinese saying " the pavilion near the water gets moon first", actually, the moon in the water is just a reflection for you to look at, it is not edible, nor useful, it definitely is not as cool as having a handsome neighbor lives across the hall from you!

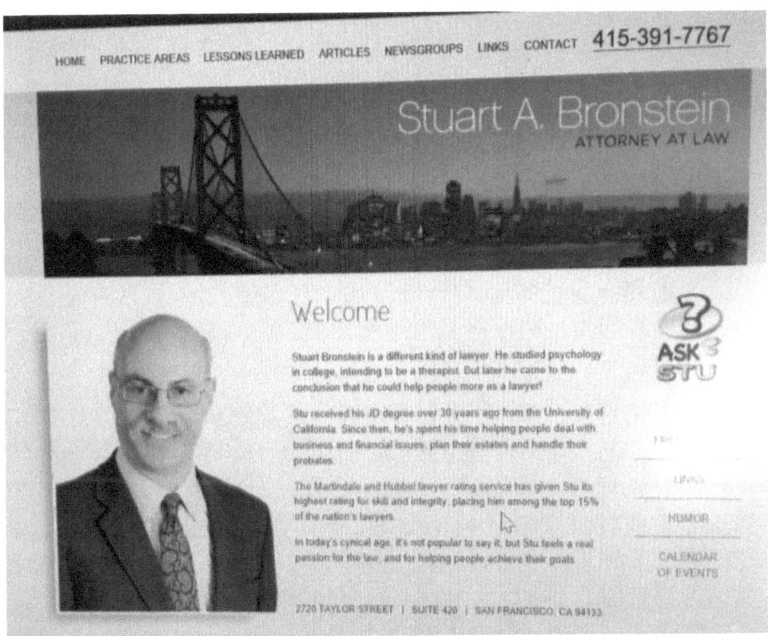

Start A. Bronstein

對門住了個帥哥

　　如玉搬進新家,第二天打開公寓的門，嚇了一大跳，你道是為什麼？原來門外站了一個帥哥，雖然看不出他禿頂上原來頭髮是什麼顏色，但是非常肯定的是他的眼睛是特別帥氣的綠色。

　　「請問…?」如玉很禮貌地問。

　　「我是住在你對面的鄰居。」那位帥哥回答。

　　「請問你有什麼事嗎？」運氣居然這麼好，對門就住了這麼一位帥哥,而且眼前就站在自家門外。

　　「哦，今天出門忘了帶鑰匙，所以回不去了！」沒有想到他的聲音也這麼好聽。

　　「這樣嗎？我可以幫你什麼忙嗎？」如玉很關心的問。

　　「謝謝妳的關心，我已經找了開鎖的鎖匠,他們不久就到了。」

　　「那…，你要不要坐下來等呢？」如玉飛快地轉身進去，搬了一張凳子出來給他坐。「那太謝謝了。」他說，就在門外坐了下來。

　　「你要不要我的電話號碼？這樣,你想叫我,打電話給我就行了。」如玉退回自己公寓之前，還不放心地對他說。說過之後，如玉覺得有點不對勁，怎麼可以隨便問人家要不要自己的電話號碼，萬一人家誤會了，如何是好？但是，一言既出，駟馬難追，問已經問了,只索罷了。

　　兩天之後，在公寓的電梯裡又遇見了這位帥哥，如玉大喜，澄清的機會終於到了。「對不起，上次我問你要不要我的電話號碼，是因為…。」話才出口，覺得更不對了，好像越描越黑嘛！

　　「這裡有我的電話，今後但凡妳有事要找我,歡迎打電話給我！」沒有想到，他不等如玉解釋，立刻由口袋裡掏出一張名片來，交給如玉。

　　如玉接過來一看，名片上不但有他的電話號碼, 415-391-7767，還有他的辦公室的地址以及郵電地址，www.DownToEarthLawyer.com 而且，原來他是一名律師呢！如玉非常高興，立刻打了一個電話給她樓上新認識的朋友:「翠花，這一下好了，我的對門是一個律師，你可以委託他替妳辦理離婚的事情喔！」

　　怎麼一回事呢？原來，如玉在電梯內遇見的新朋友翠花曾告訴如玉，她結婚還不到一年，就發現她新嫁的比她小十二歲的小鮮肉不但經常夜不歸家，就是回家也是身心不在，手機日夜廿四小時不離手，原來他不但與他已經離婚的前妻藕斷絲連，而且還天天去接送他與前妻的孩子上學，翠花找了私家偵探背後調查，照片鑿鑿，人贓俱獲，。

翠花非常氣憤地對如玉說：「哪裡是真正的愛我，明明就是想要利用我來騙取能留在美國的身分罷了。」

有美國公民身份的翠花一見到如玉，就向如玉哭訴：「現在，你說我們這樣除了離婚之外，還有什麼好辦法?」

可是，翠花和如玉都沒有離過婚，都不知道怎麼開始辦理離婚手續，現在，既然她家對面就住著一名律師帥哥，當然不能錯過這個好機會！

如此，翠花就在如玉的寓所內燒了幾個菜，由如玉按門鈴把對門的律師帥哥請過來吃中國午飯，并告訴他說想委託他代辦離婚手續。

「對不起啊！我不是辦理離婚的律師！」這位帥哥說。如玉與翠花兩人一聽，面面相覷，沒有想到律師這門職業原來分工這麼細啊！

「不過，我可以把翠花女士介紹給我那專打離婚官司的律師朋友！」他想了一下之後，一口答應。

如此這般，她們就與對門的律師帥哥打上了交道，對門帥哥的性格溫和，对人彬彬有禮，擁有很多男女朋友們 s（多數之意），除了打離婚官司的律師之外，她們通過他又認識了他的老黑房客以及其他眾多的朋友們，例如中國塑雕師、美國文人，黑人藝術家、大家都變成了朋友，到了他家就像進了一個民族大融爐，非常溫暖暇逸。

因為如玉不開車，帥哥不但開車帶如玉上街買菜，帶她一同上衙門填表格，一同吃館子，一同談天說地。

人家說近水樓台先得月，那水里的月亮只能供觀賞，既不能由水中撈起來，不能吃，更不能有什麼實用途，有什麼意思呢？哪有對門住個帥哥那麼酷呢？

Non fictional

Edited by Descartes Li

The City of San Francisco is so cold, Ruth decides to wear a little more clothes before going out to walk the dogs.

Little black dog Christine is her daughter-in-laws' dog, when the dog sees Ruth starting to put on the coat, she comes obediently to put her head into her leash, and waits patiently for Ruth to take her out. Daughter-in-law's sister-in-laws' little white dog Mandarin is not so easy to give in, it takes Ruth half an hour or so to leash her, her leash seems too big for her.

Today, San Francisco's wind is as gusty as usual, as soon as the door is opened, the cold wind starts to blow at Ruth's face, Christine takes the chance to rush to the street, then to the park, and runs on the green lawn and jumps very happily.

At this time, Ruth finds that the leashes held by her hands have different weights; the heavy one was kicked forward by Christine, and then at a closer look, Ruth can not help but to be in a panic; the little white dog Mandarin is not on the other end of her leash, she has slipped to the street as Ruth's panic-stricken eyes is watching, and then continue to run forward.

"Oh, My God, Mandarin, please do not run away!" As Ruth holds the two dog leashes, she is also shouting in Chinese to the little white dog that does not listen.

Ruth is desperate; all she is asked to do is taking care of two dogs, and now half of her charges is gone, where does the little white dog Mandarin go? How come there is no trace of her? Looking around, the end of this street is another street, should they turn left or turn right? Ruth does not know what to do, She only follows closely with Christine to run forward.

"Oh, I am going to cry!" her eyes behind the sun glasses are really filled with tears.

Wait! Wait a minute, as Ruth looks through the tears in her eyes, she sees out of the right side of the street there comes a tall and handsome figure holding a white dog in his hands, if it is not the little white dog Mandarin who else it could be?

"Lady, is this your dog?" he has a nice voice, and his blue eyes are looking at the the   two   leashes   in Ruth's hands, one has Christine at the end of the leash, the other one is empty.

"Yes, yes, the white dog is also mine!" cried Ruth.

"I'll help you to make a good rope for your dog!" The peppery haired gentleman   says kindly, and kneeling his long left leg, carefully put the white dog Mandarin in her leash.

"Oh, thank you very much, thank you very much!" Ruth was so   grateful that she was about to give him a thank you kiss.

"My name is Ivan Petrovich Ivanov." The eldly gentleman with the   Russian   name stretched out his hand, his fingers are very long, Ruth finds out that his finger nails are very clean.

"I am Chinese, and   my name is   Ruth Yan! I am so glad to meet you!" Ruth also stretched out her right   hand, very pleased that she painted her finger   nails with very popular blue nail polish today.

At this time, Mandarin jumped out of her leash again.

"Don't you worry, I will help you catch her again!" Says Mr. Ivan with assurance, and our eldly Russian gentleman turned and catchs the little white dog again.

"Mandarin   does not like her leash, I really do not know what to do with her." Ruth says to him.

"Well, where is your home? I'll escort beautiful yourself and your lovely dogs home, what do you think?" He lowered his head and asked Ruth.

"Well, please come to my house to have a cup of coffee! It is only two blocks away!" Ruth invited him with a friendly smile.

It is a pity that the two blocks are too short, and it is even more regretful that when the two of them open the door to go in, they heard that the cleaning lady Offia is   sweeping the floor.

How can any one drink romantic coffee with the deafening vacuum cleaner on?

"Then go to my house to drink coffee, my car stopped at the opposite side of the golf park." Ivan politely suggested.

Leaving   the two dogs with the clean lady Offia, they walk to the Ivan's car which is parked near the park, after a few minutes' ride, the car slipps into the Ivan's basement   parking lot, it is a newly remodeled apartment with a lift, Ruth finds out that Ivan is a head taller than herself, it is her favorite man's height.

"From the rear window of the Ivan apartment, you can see the beautiful Golden Gate Bridge amongst the mist, open the back door is the coast of San Francisco, the roadside is covered with green wild plants." Ruth tells her Son and daughter-in-law, when they come back from their vacation.

"We have to invite him over to our next weekend   party!" Ruth suggests.

"Of cause we have to! didn't he help you to find your dog back? Didn't he invite you to have coffee plus a home made Russian lunch?" her son answers.

"And, so people will know that he does exist, not as fictional as the prince from a fairy tale!"

Too bad that Ruth does not have his phone number nor his apartment address.

金門橋畔的俄國帥哥

　　舊金山這麼冷，如玉決定要穿多一點衣服才出門溜狗。

　　小黑狗克莉絲汀是媳婦的小狗，它看見如玉穿上外出衣鞋，就乖乖地走過來把頭伸進自己的狗繩，乖乖地等待外出遛彎兒。媳婦哥哥家的小白狗曼達玲卻繞來繞去，花了半小時左右,好不容易才肯就範，它的狗繩似乎不怎麼合適，好像太大了一些。

　　今天舊金山的風跟往常一樣大，大門一打開，就有一股冷風吹來,克莉絲汀一衝就衝到對街的公園里去,在綠色草坪上又跑又跳，非常歡樂。

　　正在此時，如玉就發現手中的兩條狗繩一重一輕，重的那一條是被克莉絲汀扯著往前跑，再仔細一看，不由得不驚慌起來，原來曼達玲已經不再在她的狗繩那端,那只白色的小狗已經脫開了束縛，在如玉驚慌失措的眼神中，溜到對街之後，又繼續往前飛奔。

　　「天哪，曼達玲，請妳不要再跑了！」如玉手中牽了兩條狗繩，跟在克麗絲汀後面沿著草坪跑了起來,嘴裡也在用中文大聲喊著那條不聽話的小白狗。

　　但是，不好了，小白狗曼達玲到底跑到哪一條街去了？怎麼不見她的蹤影了呢？這要怎麼辦呢？一共只要她照顧兩條小狗，才一出門就不見了一半，這可如何是好呢？放眼望去，這條街的盡頭是另外一條街，到了那裡，應該向左傳那還是向右轉？如玉真是沒主意！只好緊緊地跟隨著克莉絲汀繼續向前飛跑。

　　「還是大哭一場吧！」如玉墨鏡後面的的眼睛里果然充滿了淚珠。

　　且慢，且慢，如玉由淚眼中向外望去，只見在街道的右面轉出一個高高帥帥的身影,手中抱的那只小狗不是失蹤的小白曼達玲還是誰呢？

　　「女士，這只小狗是你的吧？」声音真好聽，他那藍色的眼睛看著如玉手中的兩條狗繩，一條上面繫著向前衝的克莉絲汀，另外一條是空的。

　　「是啊，是啊，一點也不錯，這只小狗是我的呀，只是被他掙脫繩子逃跑掉了！」如玉口中大喊！

　　「那我幫你替它把它的繩子系好吧！」這位頭髮有點花白的紳士，一面說，一面就跪下一隻長腿來，仔仔細細地把曼達玲用狗繩系好。

　　「哦，太謝謝你了，太謝謝了！」如玉心中非常感謝他，恨不得踮起腳來親吻他一下呢。

「我的名字叫做伊萬.彼得洛維奇.伊萬諾夫。」這位有著俄國名字的半老的紳士伸出手來，他的手指非常長，不知是不是因為皮膚白皙的緣故，指甲顯得非常乾淨。

「我叫顏如玉，是中國人！太高興遇見你了！」如玉也伸出自己的纖纖玉手來，非常高興她今天塗了最近非常風行的藍色的指甲油。

「又想跑了呢！」伊萬先生說。正在此時，曼達玲又蹦出去了，我們的俄國紳士轉身又一把抱住這隻白色小狗。

「曼達零不喜歡它的狗繩，真不知該把她怎麼辦？」如玉問。

「這樣好了，你家在哪啊？我乾脆護送妳這位美人和妳可愛的小狗們一同回家吧！你以為如何？」他低下頭來問如玉。

「好啊，我家就在前面兩條街道的距離，到我家去喝杯咖啡吧！」如玉笑嘻嘻的邀請他。

令人非常可惜的是，兩條街太短了，一會就到了，更令人遺憾的是，當他們兩人開門要進去的時候，聽見清潔婦娥菲雅掃地吸塵器的聲音，在樓上大声哄哄地的響著，震耳欲聾.

這怎麼能喝什麼羅曼蒂克的咖啡呢？

「那就到我家去喝咖啡好了，我的車就停在高爾夫公園的對面。」伊萬彬彬有禮地建議。

他們把兩條小狗交給清潔婦娥菲雅。兩人就邊散步邊談天到了伊萬的汽車上，只開了幾條街，車子就滑進了伊萬新式公寓的地下停車場，那是一幢有電梯的新式公寓，如玉發現伊萬比她高出一個頭，正是她喜歡的男子的高度。

「由伊萬公寓的後窗口一眼望去就看見在煙霧瀰漫中美麗的的金門大橋，打開後門出去，就是舊金山的沿海邊，路邊長滿了綠色的野生植物。」兒子和媳婦由度假回家之後，如玉把自己的艷遇告訴媳婦麗麗。

「下次家中有派對聚会，咱們非得邀請他參加不可!」如玉堅決地說。

「是啊!人家不是幫妳把小狗狗抱回來了嗎？不是請妳喝過咖啡，又請妳吃過自製俄國午餐了嗎？那我們當然要回請他!」兒子勇勇點頭同意。

「咱們一定得邀請他，不然，人家還以為他是像中國聊斎誌異裡面的書生或外國童話中的王子，是個杜撰出來的人物呢！」

可惜如玉忘了問他的電話號碼，也沒有他的住家地址。不瞞你說，這個年頭，妳要有人家的電話號碼，要有郵電地址，妳才能主動啊！

Strongwoman Scarlett Tsai

Edited by Wendy Chou

Fifteen years ago, "Little Red" Tsai and her Guangdong husband made a little money operating a street stall, selling secondhand goods in the city of Dongguan. So she sent her peasant father, who worked as a cement tile carrier, and her mother, who cooked for migrant workers, back to their hometown of Henan, and hired a temp worker to help her elderly parents to care for her autistic son.

Soon, ambitious Little Red divorced her Guangdong husband, equipped herself with a modern English name — "Scarlett" — then, as a single woman, traveled with a group of migrant Wenzhou vendors to the United States to sell inexpensive, eye-catching Chinese goods. She could spend twelve Chinese yuan on a dozen colorful scarves or 20 Chinese yuan on 12 pairs of gloves from a Dongguan wholesaler, then, at her retail booth in a flea market in California, quickly turn it into 6 US dollars per scarf or 8 US dollars for a pair of gloves. Her goods were all sought-after, and her business was booming.

One evening, after a whole day's business, a tired, hungry Scarlett Tsai dragged a large canvas bag full of small bills and change over to a food stall next to the flea market. She sat down, ordered a cheeseburger, bought a glass of Coca-Cola, and while eating and drinking, thought about finding a quiet and secure place to count the money in the big bag next to her feet.

She was exhausted, but ah! The bulging bagful of stuffed money was so enchanting! Counting it would be immensely enjoyable.

She suddenly noticed, sitting next to her, a white man who held a giant beer glass in his hand. Neck straight, he poured the beer down into his throat so quickly that he started to cough.

Scarlett saw stars and stripes on the cotton pockets of his jacket. With heavily accented English, she asked casually; "Hello, are you an American?"

The man finally stopped coughing and answered sadly: "Lady, nowadays being an American isn't easy. Look at me, I haven't found a job, so, now I've been discharged from the U.S. Army, I'm unemployed!"

"Is that so? I think it's good to be an American! If I had American citizenship, it would be so much more convenient to do my business!"

"If you want to be an American, just marry me!" He said that with a blurry tongue.

"So, what do you want from me?" Scarlett asked carefully.

"I heard that the recent market price for marrying an American for his citizenship is $60,000. I don't want more, just as much as the going rate. Or, just let me work for you, so at least I'll have a job." said the American man.

"Well, you can follow me to my booth and carry my stuffs back to my place." Scarlett suggested, thinking to herself, "This white man looks healthy enough."

Their conversation went back and forth for some time, but finally, the two really went to get a marriage certificate.

After getting married, Scarlett thought about letting her husband take part in her business. She could at least try to let him move the more cumbersome things, or be her helper.

There is a Chinese saying: "If you don't try, you will never know; after trying, you will be scared out of your skin!" This white American, named Kevin, really loved spending money. Plus, he was lazy. Apart from bed sports, he was not interested in anything. He would let Scarlett yell and scream angrily at him but remain unmoved. If Scarlet was mad at him, he would hide out and spend money in places where Scarlett couldn't find him. Often he might totally disappear.

It did not take long for Scarlett to become a U.S. citizen. She couldn't wait to divorce Kevin. Kevin, for his part, pointed out that Scarlett did not fulfill her premarital promise, and not only refused to sign the divorce certificate, but also threatened to report to the Immigration Bureau how Scarlett had married him only for his citizenship. Scarlett had no choice but to come up with the $60,000. Instead of giving the money directly to Kevin, however, she combined this money with the capital of two other Wenzhou persons, so the three could jointly invest in a winery in an affordable area of California, and rented a bed and breakfast for Kevin to manage.

Two days after divorcing Scarlett, with a winery manager's

title, the American Kevin did not hesitate to remarry. His new love, a Taiwanese woman who did not have US citizenship either, was a young and beautiful girl.

Scarlett's plan was, as one of the three investors, to periodically collect from her manager the money generated from the winery and B & B.

Unfortunately, fate does not always follow peoples' wish. When Scarlett came to collect money, Kevin handed her an extremely light money box.

"Why is there so little money? How can I pay the mortgage? Don't forget, I have to pay federal, state and county taxes?! Damn it...."   Scarlett yelled at the top of her lungs. She looked at Kevin and the Taiwanese woman clinging to him. Standing in front of her these two looked expressionless and silent...   Suddenly, she felt better, much, much better.

Ah, Strong woman! she felt a giant weight lifted from her chest!

Being a strong woman is not necessarily bad!

女強人蔡思嘉

　　十五年前，蔡小紅和她廣東老公在東莞擺地攤賣二手貨賺了一些錢，就把她抬水泥瓦片做農民工的父親和替農民工煮飯的母親送回老家河南照顧她有自閉症的兒子，還請了一個鐘點工幫助年老的父母。

　　不久，力爭上游的她又與廣東老公離了婚，自己一單身女人，改了一個洋氣的名字叫思嘉蔡，跟了一批走單幫的溫州人，到美國去專賣價廉物美的中國產品，例如在東莞十元人民幣可以批發到一打的紗巾、廿元人民幣十二雙購入的手套，運到美國加州的跳蚤市場，就變成了每條美金六元或每雙美金八元的搶手貨，生意非常火紅。

　　這天傍晚，做了一天生意，一身疲憊、飢腸轆轆的思嘉蔡，拖了一大帆布袋的美金零錢，在市場旁邊的一個食品小攤上，買了一個芝士漢堡包，叫了一杯可口可樂，一边吃喝，一边高興地盤算，要找一個安靜可靠的地方來把大袋子內的美金數一數，雖然很累了，不過袋子中的美金零錢塞得很結实，數起來一定非常過癮。

　　不意鄰座一位白人男子，手中捧了一個巨型的啤酒杯，直著脖子把啤酒朝喉嚨裡灌，因為太過猛烈，竟然咳起嗽來。

　　思嘉見他上衣的口袋是由紅色、藍色以及星條旗的布做成的，就隨便用生澀的英語問了一下：「哈嘍，你是美國人嗎？」

　　這人等咳完嗽，愁眉苦臉的回答：「女士，這年頭做美國人真不容易，你看我，一直找不到工作，所以，從美國軍隊退伍就是失業的開始！」

　　「是嗎？我覺得做美國人可好呢！我要有美國身份，做起生意來，就更方便了!」

　　「你要做美國人還不簡單，嫁給我就行了！」他大着舌頭說。

　　「那你有什麼條件呢？」思嘉小心地問。

　　「聽說最近嫁給美國人的行情是六萬美元，我不多要，只要跟市價差不多就行了。要不，讓我跟你一起做生意，至少我也有個工作。」

　　「好說，你先跟我到攤位旁去把我存在那裡的剩貨抬回去罷。」思嘉说。這白人看起來身體还不錯。

　　這樣一來二去，兩人真的去領了結婚證。

　　結婚之後，思嘉想了一下，決定先讓他一起去做生意，先試一下，至少可以讓他搬搬比較笨重的東西，也可以算做一個幫手。

不試不知道，一試嚇一跳，這個叫凱文的美國白人男子真是
爱花錢，加上懶得可以，除了床上運動以外，其他的一概是甩手
伙計，任憑思嘉如何辱罵，永遠什麼都不做，罵得兇了，变本加
厲，乾脆到思嘉見不到的地方花錢鬼混，整天不見人影。

過不多久，蔡思嘉入籍變成美國公民，忍無可忍的她立馬要
求與凱文離婚，凱文自然不願意，認為思嘉沒有履行她婚前的諾
言，不但不肯在離婚證書上簽字,并且揚言要告到移民局，說她是
貪圖他的公民身分才與他假結婚的。思嘉無奈只得把答應要付給
凱文的六萬美元頭款加上另外兩位温州人的資本，在加州土地便
宜的地方用三人的名字買了一個葡萄酒莊再加租了一個民宿交
給凱文經營管理。

老美凱文在与蔡思嘉他離婚第二天，就毫不遲疑,以酒莊經理
身分喜洋洋地與自己的新歡結婚了，他的新婚夫人是一位初由台
灣來的沒有美國公民身分的台灣女郎，長得年青貌美，甚得凱文
寵愛。

原來思嘉自以為得計，用自己的名字投資酒莊和民宿，只需
隔三差五到凱文那裡去收取營業利潤就行了，那知天不從人願，
莊主思嘉蔡過來收錢，凱文雙手捧上輕飄飄的收錢盒子。

「怎麼只有這幾個錢？怎麼夠付銀行的貸款？不要忘記，我
还得缴纳联邦、州縣等各种稅金?! 媽的，我要把這賠本的酒莊和
民宿賣了…。」思嘉愈罵愈起勁，心中一口鳥氣出了不少，再看
凱文和站在凱文身边的台湾女人兩人目無表情呆呆聽訓的像子，
心情大好。

女強人，原來做女強人的感覺十二分良好！做個女強人未必
不好！

Our Friend Bill

Edited by Descartes LI(2020/04/12)

In the city, there is a Chinese restaurant with good prices which we frequent often. The service makes people have a kind of at home feeling, not only the waiters serve the food in a neat and orderly way, but the manager is especially warm to the guests.

My son and daughter-in-law, who don't speak much Chinese, call him Bill. However, we call him "young white head" in Chinese. We call him this because even though his face looks young and active, the manager has a full head of white hair.

Whenever we don't feel like cooking at home, we go to the restaurant to eat some delicious, home-cooked food. Also whenever we have a birthday, graduation, anniversary as well as baby shower, we go to this restaurant to eat good food and celebrate the occasions together. That is to say, we view their kitchen as a kind of our own home backup kitchen. And so, over time, we got to know Young White Head pretty well.

Not infrequently, when my son took his wife there to eat supper, the manager will come over and says to them, "Hey, I read your mother's article in the World journal news yesterday!" My son and daughter-in-law would ask, "Oh! What did my mother's article say?

He would explain the contents to my son and daughter-in-law in detail.

My son would ask: "Bill, how come you remember it so clearly?"

The restaurant manager would answer with a smile, "I'm so busy every day, and have very little time to read any Chinese, but what little I do read, I remember very clearly!"

In order to look my best, at the end of last August, I went to a plastic surgeon to have a facial surgery, six month later, I was waiting anxiously for a follow up visit, when suddenly my telephone rang. It was a phone call from my surgeon's assistant, she said that in order to obey to the VOID-19 pandemic law, nobody should go out, so we have to postpone the appointment to a later date.

"Postpone it to what day?" I was startled and worried. I had some very unbecoming stitches on my face.

"I don't know, we have to wait to see which day the disease virus is under control!" And before I could completely understand what she meant, she hung up.

So what can I do? The stitches were on the top of my cheekbones. Was I really going to live my life with concave bruises on my cheeks? How long would I have to live like this? A virus is an enemy, you can't see or touch, even mediation is impossible let alone surrender! I felt so disappointed and wasn't even sure that life was worth living any more, when suddenly, a thought come to me. So many brave patients in the world struggle to survive, and so many heroic health care workers are fighting with this invisible enemies! I'm so selfish, I should be ashamed of myself!

So in order to prepare myself to stay at home for a long time, I decided that I better go to the supermarket to stock up some food. An idea suddenly hit me, and made me smile immediately. Not only will wearing a mask help prevent the spread of the virus, it can keep us from feeling cold, it can also cover the scars on my face! The facial mask can serve several purposes! I remind myself of a common Chinese saying: "One arrow, several targets."

Among all the facial masks I own, I chose a special beautiful favorite, and wear it happily to take the elevator.

Come out from the elevator, I came face-to-face with a temporary security guard of our building and he asked me, "Who are you? Where are you from? Where are you going?" wow, such profound questions, not even many great thinkers and philosophers, who have spent their lifetime pondering this, have found the answers to these questions. However, I easily resolved them in two minutes. " I'm Gwen Yu, the lady on the board of directors for this building! I'm from my apartment, and am going out to buy food."

After bought the food, I walked on the street, I saw all the shops closed tightly with their gate locked. The street, which was usually crowed with noisy shouting pedestrians, today saw only scattered, silent, mask-wearing people. One of them was a Chinese man wearing a white mask carrying some groceries towards me.

I greeted him happily from six feet. "Hello, Young white Head, Your white mask goes very well with your hair color. It looks very good on you."

"Well, thank you, last week, I went to get a dental implant, but the dentist's office was closed before putting on my dental crown, fortunately the dentist give me a facial mask to hide my missing tooth!" Wearing a white mask, he laughed and said to me aloud; "I saw your story yesterday in the World Journal News. I really loved it!   One arrow, several targets!"

"Ah, Young White Head, thank you for your kind words. You remembered the story so well!" I answered him aloud through the mask.

He answered, "As you know the restaurant is closed and I have nothing to do at home except read the newspaper. I read the same article many times, so I remember."

Keeping six feet social distance, we both laughed, long and loud.

我們的朋友比爾

　　市內有一家中國餐館，是我們常常光顧的老字號。他們的服務使人有賓至如歸的感覺，不但侍應生上菜井然有序，他們的經理對客人更是非常熱心，這位經理有著滿頭白頭髮，不過臉色很年輕，所以我們認為他是少年白。

　　因為價錢公道，平常我們自己不想燒飯了，就到他家去吃些家常菜。又因菜式可口，有誰過生日，兒子醫學院畢業、結婚、生第一個女兒和第二個兒子也是在那去慶祝。也就是說，無論大宴小酌，都到他家吃飯或宴客，久而久之，我們跟他們的經理變得非常熟悉，好像老朋友一樣了。

　　有時兒子帶了媳婦一起去吃飯，他還特地跑過來對我兒子說：「昨天在世界日報上看見你媽媽的文章！」在美國出生的兒子和媳婦就會回答說：「喔！我媽媽的文章說了些什麼？」他就一五一十細細地把內容解釋給我兒子和媳婦聽。

　　「咦，比爾，你怎麼會記得這麼清楚呢？」

　　他會笑嘻嘻地回答：「我天天太忙了，回家看中文的時間很少，只要看到，就會記得很清楚！」，唔，他的回答很有道理。

　　事後兒子的兒子和媳婦對我說說「媽媽，你把你的中文文章翻譯成英文然後由我們輪流閱讀，稍微改動一下，我負責發給所有的表哥表嫂表弟堂哥堂弟們看。」大家輪流負責，日子久了，竟然成了我們家族中一件交流心得，互通消息一件美事。

　　原是為了要美麗，去年八月廿二日，我開刀美容，經過三番二次的覆診，預定今年四月二日最後一次拆線，天天撕著日曆，數著日子，好不容易熬到三月廿八日，正在家中一面用肥皂洗手，一面心急如焚地照著鏡子，突然電話鈴響，原來是外科醫生的助理打來的電話，說是為了減少冠狀肺炎的擴散和流行，防疫期間，不得外出，所以奉命延長4月2日的預約！她說。

　　「延長到哪一天呢？」我嚇了一跳，連忙問道。

　　「不知道喔，要看哪一天疾病病毒被控制了，才能恢復工作喔！」她回答完了就掛掉電話。

　　那可怎麼辦？縫的線就在顴骨上面，真的還要頂著凹下去傷痕的雙頰活下去嗎？這種樣子還要活多久呢？有了這種看不見摸不著的敵人，連投降都不可能，更不用想要和解之前談談條件啦!正在痛不欲生的時候，突然想到世界上那麼多病人掙扎求生，又有那麼多医護人員奮不顧身，正在与這批看不見摸不到的敵人英勇戰鬥嗎？我這種只顧小我的人，太羞愧了！

　　還是到超市買點食物，以備長期在家中防疫罷！不是叫我們人人都要戴好口罩才出門嗎？保護好自己，不給社会添亂，也是善舉的一种呀！我突然靈機一動，立刻笑逐顏開，戴口罩不但防疫保暖，也可以蓋住臉上的疤痕呀！這不是一举數得嗎？

　　特地在家裡各種口罩中選了一個自己喜歡的，高高興興地戴了花色美麗的口罩，出門坐電梯去買菜也。

　　由電梯中出來，迎面遇見我們大廈的臨時保安，他對我問道：「你是誰？你從哪裡來？你要到哪裡去？」哇，這麼深奧的問題，很多大思想家、哲學家花了一生的精力鑽研，回答不了的問題，我在兩分鐘之內就輕易解決了。

　　買完菜，走到街上，但見所有的商店関閉的大門外還有鉄柵緊緊地鎖住，平常熙熙嚷嚷，行人如織的路上，只剩稀稀朗朗，寥寥數人。一眼就遠遠看見一位提了菜蔬，帶了白色口罩的華人在對街与一位白人撞肘、碰拳之後，走了過來，大概餐館奉命暫時停業，他才有機会上街買菜。

　　「哈嘍，比爾，你好！你的白色口罩跟你的頭髮顏色很配，非常好看。」我在六呎以外的距離，苦中作樂地跟他打招呼。

　　「是嗎，謝謝妳，其實，我去種牙，只種了牙根，還沒有安上牙冠，牙醫診所就暫時不上班了，幸好牙醫送了我一個口罩擋住我的缺牙！」戴了白色口罩的他笑了起來，大聲地對我說：「我昨天在一本中文的美食雜誌上，看見你的大作了！」

　　「嗨，比爾，謝謝你有心還記得告訴我！」透過口罩，我我也大聲地回答他。

　　「最近餐館不營業，我在家中枯坐無聊，一份中文報紙雜誌看上好幾遍，所以你的文章我都記得了。」

　　走在六呎以外的我們，遠遠地無可奈何地大聲笑了起來。

A Diamond Ring

Edited by Descartes Li

A Diamond Ring

More than 20 years ago, Jade received a phone call from her husband Tyson's distant cousin Guangson; "Jade, I am in the United States on business for this weekend. Is it ok for me to stay at your house?" Guangson asked.

Guangson owned an import/export company which shipped merchandises from China and sold in United States. Although he came to the US very frequently, but he was so busy that he seldom had any chance to visit with Tyson and Jade.

"You are very welcome!" Jade answered happily.

Sure enough, that weekend, they picked up cousin Guangson at JFK airport, and then drove him to their home on Long Island, New York.

Guangson told them that he was going to retire soon, and, did not know when he would be back to the United States again. He wanted to come to Long Island to see how Jade and Tyson enjoyed their retired life, and, in addition to that, he also wanted to buy a diamond ring for his wife Baili before his retirement.

"We have married so many years. Baili has been taking care of my mother, her own mother, plus our three children who are going to school back in Hong Kong. Baili stays in Hong Kong where she is busy taking care of all of us, old and young, day and night. It is really not easy for her! So I want to buy her a diamond ring with the money I've saved over the years! I heard that in the United States you can buy a better quality diamond ring with the same amount of money.

Would you be willing to help me out?" Guangson said with great sincerity.

Guangson and Baili were married in Yunnan where they met. They were "educated youth going to the countryside" during the Cultural Revolution. Baili looks like her ethnic minority Bai mother: very beautiful with light complexion. However, her personality was along the lines of her Han father: hard-working, gentle and quiet.

Although Jade and Tyson were no diamond ring experts, they still very enthusiastically took Guangson to a famous jewelry shop in Long Island, New York. They helped him to buy a diamond ring worth about 10,000 dollars, and then escorted a happy cousin Guangson back on to the plane to Hong Kong.

Unexpectedly, about one month later, they received an emergency long-distance call from Baili in Hong Kong, she said that since Guangson was in Dongguan doing the company business that weekend, she would fly to the United States to Jade's house.

A very panicky Baili arrived Long Island, and she immediately asked Jade and Tyson to take her to the same jewelry store where her husband Guangson bought the diamond ring. She wanted to purchase an identical ring.

What was going on? It happened that one day, Baili was in the Hong Kong restaurant drinking tea with her lady friends. She could not help but show off the diamond ring that her husband had so gratefully given her. She took off the ring from her finger for the ladies to pass around to admire, but it never came back! She was so upset that she eventually she took out 10,000 dollars of her own, money that she had saved for a lifetime, flew to New York, and was now back at Jade and Tyson's home, telling this story.

Baili didn't sleep well that night. The next day Jade and Tyson took her to the very same jewelry store, but the clerk told them that the ring she wanted had been sold out.

"This one, this one is a bit like it!" At a glance, Baili saw another diamond ring inside the glass counter and yelled with joy.

"Just like it? Not exactly the same? What if it's recognized by Guangson?" Jade was unsure.

"It doesn't matter. Guangson will never be able to recognize it!" Baili was very confident of Guangson's carelessness.

That very evening, Baili flew back to Hong Kong in a hurry.

20 years later, Guangson took Baili to travel all over the world. As part of this trip, they went to Florida to visit their retired cousins Jade and Tyson. When they entered the door, Jade and Tyson saw Guangson was   holding his wife's leather shoes with one hand, the other hand was wrapped around his wife's waist. The diamond ring on Baili's finger was quietly shining with a warm and soft light.

Seeing the sweet couple so much in love, Jade and Tyson's hearts beat more comfortably.

After the lovely visit from Guangson and Baili left, Jade and Tyson had a little chat between themselves.

Jade could not resist her curiosity, so she asked her husband Tyson. "Hey, it seems that Guangson didn't find out that the ring on Baili's finger was bought with her own pocket money, or did he?"

Tyson laughed. "Of course he did!"

"What?!" Jade was startled.

"Guangson told me that one day he saw two identical jewelry boxes on the Baili's dresser. The receipts inside had different dates, so he had his doubts. Initially, he thought that Baili had asked us mail order it. Later, he learned from me that Baili personally flew to the United States to purchase the new ring. Of course, Guangson was so moved by his wife's dedication!"

"Will he tell her?" Jade asked again.

"Guangson intends to publicly confess to her at their 50th anniversary party. Then he'll give Baili a pair of matching earrings that he bought secretly while they were visiting Long Island!"

一枚鑽戒

夫妻相處之道

二十多年前，如玉收到老公台生的遠房堂弟廣生來的一通電話。

「四嫂，這次我到美國，在紐約長島你家去住一個週末，不知方便不方便？」廣生公司平常的任務就是將中國生產的貨品由東莞運到美國邁阿密去。雖然常常到美國來，但卻因為忙碌的關係不曾到堂哥台生家住過。

「歡迎，歡迎！」如玉非常高興的回答。

果然那個週末，他們由甘納地機場，將廣生接到長島他家。來了之後才知道原來廣生因為退休的年紀快到了，不知什麼時候才會再到美國來，想來長島看看台生夫婦退休以後的生活情況，另外，他想在退休以前買一枚鑽石戒指送給他的太太白麗。

「結婚這麼多年，白麗因為要照顧我的媽媽也就是她的婆婆以及要照顧她自己的媽媽，加上三個孩子都在香港讀書，就一直住在香港，她為了我們這些老老少少，一天到晚忙個不停，真是難為她了！所以我想將這麼多年來我儲存的零用錢替她買一枚鑽石戒指！聽說在美國購買比較貨真價實一些，所以想請你們幫忙。」廣生非常動情地告訴台生夫婦。

廣生與白麗倆人是在當年知青上山下鄉時代在雲南鄉下相識而結婚的，白麗長得像她白族的媽媽，非常漂亮。性格卻像她漢族的爸爸，吃苦耐勞，溫順而安靜。

如玉和台生雖然對鑽石戒指也不怎麼內行，但卻義不容辭地帶了廣生到紐約長島一家比較有名氣的珠寶行選購了一枚價值約一萬元美金的鑽戒，然後護送堂弟廣生歡歡喜喜地坐上回香港的飛機。

不料大約一個月以後，竟然收到了一通白麗由香港打來的緊急長途電話，說要要趁廣生到東莞去處裡公司事務的那個週末，到美國如玉家來一趟。神情非常慌張的白麗來長島之後，立刻要求台生夫婦帶她到廣生買鑽戒的同一家珠寶行去再買一枚一模一樣的鑽戒。

怎麼一回事呢？原來白麗為了顯擺老公廣生送她的戒指，與她的閨蜜們在香港酒樓飲茶的時候，不合把鑽戒由手指上摘下來讓閨蜜們傳觀，那知傳出去之後就沒有再傳回來，這一下白麗魂飛天外，幾乎不想活了，思前想後終於找出了自己一輩子存下來約一萬元美金的零用錢，想出這一招。

風塵僕僕的白麗一夜沒有睡好，第二天台生夫婦帶她到珠寶行，不意那家店員卻告訴他們說她要的那一款戒指已經賣光了，如玉一聽，一顆心立刻沈入谷底。

「這一枚，這一枚有點像！」白麗一眼看見玻璃櫃檯裡面的另一枚鑽戒，歡呼了起來。

「只是像而已嗎？不是完全一樣？萬一被廣生認出來了呢？」如玉真的不放心。

「不要緊，粗心的廣生一定認不出來！」對這一點，白麗非常有信心。買了新鑽戒的當天晚上，她就急急忙忙地趕回香港去了。

20 年後，廣生帶了白麗到世界各處去旅遊，兩人特地到佛羅里達來探望退休了很久的表哥和表嫂，進門的時候，只見廣生一隻手提着老婆的皮鞋，另一隻手摟著老婆的腰，白麗手指上帶著的那枚鑽戒安靜恬蜜地發著柔和溫馨的亮光。如玉和台生見他们兩人這樣甜蜜恩愛，心裡非常安慰。

開車送走了廣生和白麗，事後台生夫婦兩人私下討論。

「喂，看來廣生沒有發現白麗戴的是她用自己存儲的零用錢購買的鑽戒吧？」如玉不放心地問台生。

「當然有！」台生笑道。

「什麼?！」如玉吓了一跳。

「廣生告訴我說他見到白麗梳妝台上兩隻一模一樣的珠宝盒，裡面的收據卻是不同的日期，就心存疑問了，他當初还以為是白麗請我們郵購的呢!後來知道了是白麗親自到美國來選購的，當然格外感動了!」

「他不會告訴她罷?！」如玉又問。

「白麗最近忙着籌備慶祝他們五十週年的金婚念，廣生打算在他们金婚記念日當眾向她告白，再加送他這次瞞着白丽又在同一家珠宝行購買的一對耳環呢!」

Edited by Descartes Li

As usual, Jade got up around six in the morning and pounded away at the keyboard on her computer. Did she hear the doorbell ring? How could it be? It was so early! She shook her head, she must hear it wrong.

When she was about to start working again, the doorbell rang again.

She went to open the door to take a look, it turned out that Bella Yu was at the door.

"How come you don't answer my message?" Bella was very angry.

"What message?" Jade was confused, she could not understand what was going on.

"Just take a look at your cell phone!" Bella was in a bigger rage.

Jade took a look at her phone, Indeed, Bella did call her several times at four o'clock in the morning, unfortunately, Jade often turned off her phone at four o'clock in the morning.

"Okay, all right! Have a cup of your favorite coffee! Do you like to have some sweets to go with your coffee?" As Jade finding a cup to make a cup of hot coffee, she asked.

"Huh! I don't want to be as fat as you are!" Bella pulled the skirt on her slender waist.

When Jade heard that, she was a little ashamed of herself, she really has put on too much weight lately!

"since you've made the coffee I like, I'll just forgive you again!"

"What's there to forgive me? Are you kidding?" Jade startled.

"It was not funny! When did you see me joking? I'm not kidding!" No, Bella was right, she had never made any joke.

"Ok, Then you're welcome to stay here for coffee, and I'll go in to finish the work I was working on!" Jade answered.

"It's not polite; the guest is here and you don't keep your guest company!" Bella pouted her beautiful little mouth again.

Jade felt Bella was not entirely unreasonable, so she stayed, and patiently listened to the Bella's complains.

"Jade, How come you are so dark? No wonder you can only have a darkie boyfriend!"

"Bella, as long as the persons' heart is kind, skin color is not important!"   Jade retorted.

"Dark skin is ugly! Look how white I am! "

"Is it so? I'm a senior black beauty then! " Jade laughed with good temper, She really appreciated her own humor.

"Ah Ya! my God! Beautifull or not was not said by oneself, a person can not call herself beautiful, just look at me, I was praised since my young age, everybody says that I am beautiful, that I look like a movie star, but I, myself have never said that I am beautiful!" Jade did not know why Bella was so serious.

Hearing Bella said so, Jade could not help to open her mouth, when she was about to say something, but, after carefully think twice, why bother, she decided to shut up and keep silent.

"You look at me, my married white man husband Larry sends $10,000 a month as my pocket money for me to spend, think about that, ten thousand just pocket money! He lives away in Wisconsin, we don't live together, so, my pocket money doesn't include his own living expenses...." Bella said.

"It's just, why does he have to work for a funeral homes? Why not find a better job? He is so poor!" It took a while for Jade to realize Bella was talking about her black friend.

"Bella, I do know that a person's character could be low or noble, But, no occupation is lower or less nobler than any other occupation!" Upon this, Jade, who had intended to keep silent, could not help but start defending her friend again.

"What do you mean by No cheap occupation? Look at Tea Olive....." Beginning   to criticize their mutual friend Tea Olive, Bella started to gabbling non-stop.

"Bella, our sage have three rules; hears no evil, sees no evil and speaks no-evil...." Jade said to Bella.

"Apologize, you have to apologize to me, or I'll never speak to you again!" To Jades' surprise, Bella shouted at the top of her lung, suddenly, she stood up and rushed to the door and left the coffee cup unwashed.

"Hum! Those were said by the ancient sages, if you need apology, go and find them to apologize to you!" Jade heaved a sign of huge relief, thankful that she will be able to finish her work which she has promised her editor.

Three days later, Jade went to the beauty shop to have her hair cut, the beautician told her, after she cut her hair please go upstairs, there is a lady waiting for her.

This made Jade curious, did not know who was the lady upstairs waiting for her.

"Say sorry to me, you owe me an apology....." The mysterious person turned out to be our modern princess Bella! Because she does not love exercise, so she needs someone to massage her to help her blood circulation.

Since Jade had entirely forgotten the previous incidence, so she couldn't help asked; "Otherwise?"   As Jade was asking, she reflected on the whole thing, perhaps she was really not polite enough.

"My white gentleman Larry is flying down from Wisconsin to California to see me in two days,....." Bella announced triumphantly.

Good for her! Thank god that the beautiful princess does have a white gentleman husband, he must be the most patient man who could put up with our modern princess Bella, Jade could not help but to admire him! But, since they loved each other enough to get married, why not living together? That is, one of them lives in Wisconsin and the other one lives in California?

現代公主尤美麗

　　與往常一樣，如玉清晨六點左右起床，在電腦上操作鍵盤。

　　怎麼好像聽見門鈴響。怎麼可能？這麼早！她搖了搖頭，大概聽錯了。當她正想要重新開始工作的時候，門鈴又響了起來。

　　開門一看，原來是尤美麗。

　　「怎麼你都不回答訊息的？」美麗非常生氣。

　　「什麼訊息？」如玉還不能進入狀況。

　　「看你的電話就知道了！」美麗愈發怒道。穿了高跟鞋的腳踏進公寓，她到如玉家，是從來不脫鞋的。如玉打開自己的手機，原來美麗清晨四點果然是打了好幾通電話給她，可惜清晨四點如玉的電話經常是關機的。

　　「好啦，好啦！喝杯妳喜歡的咖啡吧！要吃些什麼來配咖啡嗎?」如玉息事寧人地說，找出咖啡杯來泡了一杯熱騰騰的咖啡。

　　「哼！我才不要像妳這麼胖哩!」美麗扯了扯自己纖腰上的裙子。

　　如玉一聽，心裡有點慚愧，自己近來實在是太胖了!

　　「不過看你泡了我喜歡喝的這種咖啡，就勉強再原諒妳一次吧！」

　　「有什麼好原諒我的呢？你是開玩笑吧？」如玉嚇了一跳。

　　「有什麼好開玩笑的？你什麼時候見我開玩笑的呢？我才不開玩笑呢！」是喔，美麗是從不開玩笑的。

　　「哦，是這樣嗎？那你在這裡喝咖啡，我進去把正在做的工作完成罷！」如玉說。

　　「真是沒有禮貌，客人來了也不陪客人！」美麗又噘起她美麗的小嘴。

　　如玉一聽，覺得美麗也有她的道理，只好依然坐在原地，聆聽美麗埋怨。因為美麗真的從不開玩笑的。

　　「如玉，看你這麼黑，難怪你只能有一個黑鬼男朋友！」

　　「美麗，人只要心地善良，皮膚什麼顏色不重要的！」如玉反駁。

　　「黑就是長得醜啊！你看我多白！」

　　「是嗎？我是資深黑美女呀！」如玉好脾氣地笑了起來，非常欣賞自己的幽默。

「哎呀！我的老天！美女不是自己叫的，一個人不可以自己叫自己美女，妳看我，從小都是人家叫我美女，說我長得像電影明星，可是我從來不說自己是美人！」沒有想到美麗居然這麼認真。如玉聽她這麼說，張開了嘴，忍不住又要說什麼，但是，仔細想了一下，何必多費口舌呢，決定閉口不再說話。

「妳看我，我嫁的白人先生每個月讓我花一萬元美金做零用，只是零用錢喔，他住在威斯康辛，我們並不住在一起，我的零花錢並不包括他自己的生活費用……。」美麗說。

「只是，黑鬼為什麼在殯儀館工作？怎麼不找一個好一點的工作呢？所以這麼窮！」美麗繼續說。說了好一陣子如玉才了解美麗原來說的是如玉的黑人男友。

「美麗，要知道世界上人品有貴賤，職業是沒有貴賤的！」本來打算要不講話的如玉，忍不住又開始辯解。

「什麼叫做職業沒有貴賤？你看桂花……。」美麗開始批評她們共同的朋友桂花，濤濤不絕了廿分鐘左右。

「美麗，咱們聖賢有三句話，叫做非禮莫聽，非禮勿視，非禮不言…。」如玉實在忍不住，正打算站起身來。

「道歉，你得向我道歉，不然我不理你了！」美麗突然大喊，站起身來向門外衝去！留下未洗的咖啡杯。

「喔！這可是聖賢說的話，妳要道歉，找古聖先賢們向妳道歉罷！」正好，不必再聽這些，可以進房把答應人家的稿子完成，如玉鬆了一口大氣。

三天以後，如玉到理髮店去剪頭髮，理髮師告訴她，剪完頭髮請到樓上美容院去，有一位女士在等她。這使如玉有點好奇，不知一位什麼樣的女士在樓上等她。

「抱歉，你得向我道歉……。」在樓上美容的原來是公主尤美麗，對了，因為她不愛運動，只得找人按摩通經活血。而如玉呢，早就不記得為什麼要道歉了。

「不然呢？」如玉忍不住問，并且同時反省了一下，可能自己真的不夠有禮貌，她們才認識不久。

「我的白人先生恩理，過兩天就要來看我了，……。」公主美麗宣佈。

謝天謝地，這位美麗的公主還有一位白人先生附馬爺，能夠做這位公主的駙馬，修養一定是一級棒的了！真令人不得不佩服他！只是，他們既然情投意合結了婚，為什麼一個住加州另一位住威州呢？

---edited by Sandra Chin

Visiting my parents on weekends better be early because they get up early.

I pushed open their door carrying in the chocolate cream cake. Mom and my younger sister Katherine were eating rice porridge with bits of preserved radish. Across from them at the table, Dad was eyeing a glass jar filled with a wide range of sandwich biscuits.

"Denise, I'm brewing a pot of freshly ground coffee this morning and trying to decide which kind of biscuit goes better with it," he said as he got up, came to the door, took the cake from my hand, and then went to find me a cup for my coffee. It looked like Dad had already tried a lot of biscuits with his coffee.

Wow, I couldn't help but take a deep breath. The kitchen was filled with the aromatic smell of coffee.

"Dad, it's us fat people with big hearts and wide bodies who are as stable and reliable as huge rocks," I said, as I cut two large pieces of cake, put them on plates, and then sat down. Here we were, father and daughter, each one with a giant plateful of goodies.

"Eating too many calories isn't good for your heart!" Katherine pronounced as she finished her porridge and started to clean up her side of the table while eyeing our plates.

"After we eat all these, Dad will take his medicine that lowers his blood pressure," I pleaded as if bargaining with her.

"Of course, one must take one's medicine seriously," she answered. "Better yet, mom and dad should go to the senior citizen center and make friends their age to share activities that are good for their hearts and bones." Then, following Mom, she carried the bowls as mother and daughter went into the kitchen together to clean up the dishes.

"The senior citizen center has a lot more ladies than gentlemen. The ratio is too steep," Father said hesitantly.

"Dad, isn't there a Chinese saying to describe the situation: 'one red flower among ten thousand green leaves!' Don't forget, the rarer something is, the more precious it is!" I laughed out loud and commented.

"It's not that simple!" said Dad.

Before Dad finished, I saw my sister open the kitchen door and come out.

"No one needs an injection? No!? Then I am going back to my own home." Katherine put on her short, white doctor jacket and left.

A few days later, I was in the laboratory immersed in an experiment when my research assistant rushed in saying I had an emergency call from an elder lady. My heart immediately missed a big beat and my ears started to ring, as if the Buddha had hit my head with a Buddhist rod.   In a panic, I dropped off the test tubes, rushed back to my office, and answered the phone.

"You ... bull shit!" The voice was sharp and high.

"What?" I thought my mother was yelling at me in English. How could it be?

"I'm looking for my daughter, Dr. Yu," mom stammered in English. Oh! She wasn't yelling at me, she was looking for her daughter!   "PhD doctor" in Mandarin does sound like a curse in English.

"Mom, this is your daughter Denise!" My voice started to shake, and my heart started beating like an earthquake had exploded in my chest.   My mother had never called me at my laboratory before.

"Ah, oh, ah, please come as fast as you can, your father is not....right!" I could hear Mom shout as her voice cracked.

There was no time to take off my lab coat as I flew to the parking lot.   I wept as I started the car engine and couldn't stop even after I arrived at my parents' house.   Mom was still crying when she came to open the door for me.   Her hair was uncombed and her face was as sallow as yellow wax. Before I exchanged any words with Mom, I rushed into the bedroom and saw my father lying in their bed, moaning loudly that his whole body hurt, and that he wasn't able to get up from his bed to receive me.

Soon, my sister in her doctor's uniform stumbled into the door.

"Is it a high blood pressure stroke?" My sister burst into

tears and cried loudly.

"This kind of sickness is extremely terrible! Yesterday, your dad was able to sing and dance like a normal person. Then after one night's sleep, he woke up this morning and his whole body is covered in pain." Mom was drowning in her tears.

My sister fought back her grief, approached the bed, looked at dad's tongue, and then checked his pulse.

"How is Dad?" Mom and I asked, holding our breaths.

"I can't find anything wrong. For now, I'll prescribe some painkillers and take dad to the hospital for a thorough exam," Katherine sobbed and sniffled, as a string of bean-sized tears dropped on her stethoscope.

I went back to my laboratory full of fear and nightmares.

After one, then two days without any news, I couldn't help but call my mother to ask, "How is dad?"

"Oh, the diagnosis came. It said that your dad danced too much so there was too much lactic acid in his muscles. That's what caused him so much pain!" Mother replied.

"What can I do? The senior citizen center has more women than men so each man has to dance ten men's share. Who am I to have enough guts to refuse my responsibility?" I heard Dad complain faintly through the receiver.

"I told you a long time ago, it's not a simple task surviving in a group of Yingyingyanyan (beautiful women). It's so much easier said than done!"

萬綠叢中一點紅

週末去老爸老媽那兒，早些比較好，他們都早起。

我興致匆匆地提了巧克力奶油蛋糕進他們門時，廚房中老媽與小妹已經喝著和稀飯配蘿蔔乾。餐桌對面，老爸雙眼聚精會神地看的是一個玻璃罐子，罐中裝的是各式各樣的夾心餅乾。

「阿英，我今天煮了一壺現磨咖啡等妳，現在正在嘗試看配哪一種餅乾比較合適。」他老人家看見我推門進來，慌忙接過蛋糕，起身去替我找咖啡杯子，看樣子老爸早已嘗試了不少餅乾配咖啡。

哇塞，我不由得深深地吸了一口氣，廚房裏到處瀰漫著咖啡的香味。

「老爸，還是我們這些心廣體胖的人，又安泰、又穩重，好像磐石一般的可靠。」我一面說，一面切了兩大塊蛋糕，用碟子盛了端到老爸面前坐下來，父女二人，一人一大碟。

「吃太多卡路里對心臟不好呀！」小妹將最後一口稀飯喝完，站起來收拾碗筷時，眼睛一直對我們的蛋糕瞄著。「吃完這些，老爸馬上就吃上一次妳開給他減低血壓的藥啦！」我說。聲音裏透著討好，又像是與她在討價還價。

「當然要按時吃藥。最好，還是到老人中心找與爸媽相同年齡的朋友們做些活動筋骨的運動才是可。」說完，她就端了盤碗，跟在老媽身會後，母女一同進廚房裏去清洗餐去了。

「老人中心女多男少，比例太懸殊．．．。」老爸猶豫道。

「那不正是所謂的；萬綠叢中一點紅嗎？不要忘記，物以稀為貴呀！」我笑著接道。「哪裏那麼簡單！．．．。」老爸說。老爸還沒有說完話，只見妹妹推開廚房門走了出來。「有沒有人需要打針？沒有的話，我就回自己家了。」小妹一面套上她的醫師短白外套，一面朝外走。

過了幾天，正在實驗室裏埋頭專心做實驗，我的研究助理突然跑來說有一個緊急電話，一聽，我我的耳朵馬上轟的一響，好像當頭棒喝，急急丟下手中的試管試罐，慌慌張張地回到辦公室去接電話。

「妳．．．Bull Shit！」媽媽的聲音，尖銳而且高昂。

「什麼？」我以為媽媽用英文罵人，怎麼可能呢？

「我找我的女兒打狗拖李。」媽媽的英文夾著中文，結結巴巴。喔！不是罵我，原來她是要找李博士。

「媽媽，是我，是你的女兒阿英！」我的聲音發起抖來，一顆心跳得驚天動地的響，媽媽從來不曾打電話到我實驗室來過。

「阿英呀，妳快來呀，你的爸爸不得了了啦！」媽媽喊道，聲音極其淒厲。

我來不及換下長實驗衣就往停車場飛奔，我那不爭氣的眼淚，由汽車引擎發動起，一直滴到爸媽的家門。

媽媽哭得頭髮散亂，臉色蠟黃地來開門，我搶進臥室一看，爸爸躺在床上大聲的呻吟，說是渾身疼痛，爬不起來了。

不一會兒，妹妹穿了醫師的白短制服，也哭哭啼啼，跌跌撞撞地衝進門來。

「不是高血壓中風吧？」妹妹放聲大哭。

「這種急病真是太可怕了，昨天還好好的，可以唱歌也可以跳舞，不知怎麼睡了一夜覺，今天醒來就渾身都在疼。」媽媽早已哭成淚人兒了。

妹妹強忍悲痛，走近爸爸床前，先看舌苔，再按脈博。

「怎麼樣？」媽媽說我同時屏氣凝神問道。

「查不出來，只得先吃點止痛藥，然後我帶爸爸去做全身檢查罷。」妹妹抽抽噎噎地說，一顆顆豆大的淚珠，滴滴答答地掉在聽針筒上。

我只得提心吊膽地再回到實驗室去。

一天兩天沒有音訊，我實在忍不住了，打了一個電話給老媽，問道：「老爸怎麼樣？」

「喔，診斷書來了，說老爸跳舞太多，筋肉裏的乳酸太濃，才會這麼痛的！」媽媽回答道。

「那有什麼辦法，老人中心女多於男，一個男人要當十個男人來跳舞，怎麼好意思不勉為其難呢？」電話裏我聽見老爸在說。

「早就告訴過妳們，在一群鶯鶯燕燕裏面生存，談何容易！」電話那端老爸的聲音還在遠遠地抗議。

The handsome Cantonese guy in San Francisco

Edited by Descartes Li

As soon as she moved to California, Ruth remembered when her parents lived in California, At the time the two elderly people happily went out early in the morning and come home safely in the afternoon was due to their participation in various great activities of the senior Center, that's why their life was so colorful.

With this in her mind, Ruth checked on the Internet to see which senior center in San Francisco she can attend, and sure enough, on Geary street, not far from her son's home, there is a senior center, and she also find out that there is a direct bus right from her son's home to get there, every Tuesday they serve free lunch, can it be any better?

On Tuesday noon, Ruth changed into a new dress and had a fancy hairdo, got on the bus, it only took her ten minutes to reach the so-called [Leisure Center], and sure enough, the bus stops right in front of the building.

Ruth walked into the entrance, behind the counter there sat two ladies speaking Cantonese to each other.

"Excuse me, do you have any free lunches for the seniors here?" Ruth smiled as she asking in Mandarin with a Taiwanese accent.

The two female staffs looked at each other, they did hear Ruth's Mandarin but did not understand it! The three exchanged some gestures for a while, like chickens talked to the ducks, no result.

At this moment, from outside, there came in an elderly guy with a peaked cap and a black jacket. After listening to their conversations, he said in Mandarin with a lovely Cantonese accent to Ruth: "This is San Francisco Self-help Jackie Chan Leisure Center, one has be a member here to enjoy free lunch, Do you want to be a member there? There are some blank forms on the shelf, you can fill them out, and become a member! " After that, the peppery haired gentleman Not only took the trouble to walk past them to take the form, but also from his jacket pocket took out a ball pointed pen for Ruth to use.

"You are now a member of this association, please come in to enjoy a free lunch! ... However, although it is free, but do hope you can donate two dollars." Handsome guy explained.

Ruth took out   two piece of one dollar bill from her purse and gave them to the staff, and then looked around,   and at one glance out of the corner of her eye, she saw an vacancy at a far away corner table, she hurriedly walked there and happily took   that empty seat.

Unexpectedly, the handsome guy also paid two dollars, followed her into the cafeteria, and stood in front of Ruth, his eyes have been watching Ruth for a while, and then Ruth found, that all the old gentlemen at their table also looked at her, Ruth was confused also was a bit embarrassed, Is it that the dress she wore today was extremely pretty? Or, her hairstyle today is particularly stylish?

"Hi, lady, you're sitting in Tony's chair!" Someone next to Ruth said.

"Oh, did I take up his place!?" Ruth suddenly realized, "No wonder he and his friends were staring at me!" the color of Ruth's face has turned from pink to red, She did not know should she stand up and give up the seat? Or should she continued to bow her face and sat there blushing?

Shortly after the stalemate was maintained, there came the announcement of the lunch was served. The old Cantonese gentleman went to the office and dragged another chair out to sit next to Ruth, Then, everyone began to eat.

"Tony, I'm so sorry, it's your place!" In the middle of high and low, large and small Cantonese, Ruth apologized to the handsome Cantonese with Mandarin with Taiwanese accent.

"It does not matter, am I sitting here very comfortable? Is today's   cauliflower delicious?" Tony asked Ruth.

"Yes, Is this   Cantonese style cooking?" Ruth asked.

"How could these be counted as Cantonese flavor! give me your number, and next Tuesday, I'll take you to the delicious Cantonese restaurant in San Francisco to taste the authentic Cantonese cooking!" Sure enough, next Tuesday, Tony Introduced Ruth to a famous Cantonese restaurant in San Francisco to eat a very delicate and delicious Cantonese dim sum.

Tony told Ruth that when he was young, he graduated from Guangzhou Jinan University and came to Los Angeles, where he brought a large number of   Mexican workers to grow edible mushrooms in the abandoned mine pit and met a Taiwanese girl, and they got married very happily, but unfortunately it   did not last, his wife died after his retirement, they had no children.

"My late wife is a Taiwanese food expert so I know some of the best Taiwanese restaurants in San Francisco too, and I'll take you someday to try it!" So they later went to a few better Taiwanese and Northern restaurants.

"The handsome Cantonese guy is not bad at all, but he has suggested to rent an apartment to live with me, probably he is not that well to do." As Ruth talked to her son and daughter-in-law commented.

"Grandma, where did you meet this gentleman?" Ruth's daughter-in-law inquired.

"I met him at the luncheon at the Jackie Chan Recreation Center, an elderly self-help office in San Francisco," answered Ruth.

"How much is the lunch there cost?" Her son's wife asked.

"The   donation is US $ 2 per person," answered the grandmother.

"Look! grandma, would you prefer to meet a cute, good-looking elderly gentleman? Or to meet some body loaded with money? If you want to find the latter, you should have found a place where the they charged twenty dollars, or even two hundred dollars for lunch. " The American daughter-in-law gave an good advice to her Chinese mother-in-law.

www.amazon.com/author/gwen.li

舊金山的廣東帥哥

　　一搬到加州，如玉就想起當年她的父母在加州住的時候，兩位老人家每天一大早就高高興興地出門，直到下午才平平安安的回家，就是因為參加了老人中心的各種活動，所以生活才那麼多姿多彩的。

　　這樣一想，如玉特地上網去查了一下，看看舊金山有什麼老人中心她可以參加，果然，在距兒子家不遠的地方，就有一所老人中心，再一看，原來從兒子家的門外就可以搭公共汽車直達，每個禮拜二都有免費午餐可以吃，那豈不是更好了！

　　所以，星期二快到中午的時候，如玉就穿了一件自以為花俏年輕的衣服，坐上公車，才十幾分鐘就到達了這個所謂的［康樂中心］，果然，這個中心就在汽車站旁邊。

　　如玉喜滋滋地走了進去，進口處有兩位女士坐在櫃台邊用廣東話談天。

　　「請問，你們這裡有給老人吃的免費午餐嗎？」如玉笑嘻嘻地用台灣國語問道。

　　那兩位女工作人員面面相覷，對如玉的國語有聽沒有懂啦！三人比劃了半天，還是雞同鴨講，沒有結果。

　　正在此時，外面走進來一個頭戴鴨舌帽，身穿黑夾克的一位老年帥哥，他聽了三人的對話，非常和顏悅色地用一種可愛的廣東國語對如玉說：「這裡是舊金山安老自助處成龍康樂中心，一定要是會員，才可以享受免費午餐，妳想做會員嗎？那邊架子上有一些表格，你先填寫一下，就可以成為會員了！」說完後，這位老先生不但不厭其煩地特地走過去把表格拿過來，還由自己夾克的口袋中掏出一支筆來，讓如玉使用。

　　「妳現在已經是本會會員了，請進去享受免費午餐！…，不過，雖然是免費，但是還希望你能樂捐兩元美金。」帥哥解釋道。

　　如玉由皮包里取出兩張一元的美金交給工作人員，然後就抬眼朝裡面看，一眼望見遠遠一張桌子上有一個空位，她連忙三腳兩步趕了過去，高高興興地坐在那空位置上。

　　沒想到，這位好心的帥哥也付了兩塊美金之後，跟著她也進入餐廳來了，而且站在如玉面前，雙眼一直看著如玉，看了一陣子，如玉這才發現，全桌的老先生也都朝她看，弄得如玉都有點不好意思了，難道是今天穿的這件衣服非常好看嗎？還是今天的髮型特別時髦呢？

　　「這位女士，你坐了湯尼的位置啊！」旁邊有人告訴如玉說。

「哦，原來我佔了他的位置！」如玉恍然大悟，難怪他和大家一直看著我！如玉的粉臉一直紅到耳根，一時之間，不知道自己是應該站起來讓位呢？還是一直紅著臉，低著頭坐在那裡。

僵局維持了不久，裡面傳出要開飯的聲音，那位廣東老先生才走到辦公室裡面去拖了另一張活動椅子過來，坐在如玉的對面，大家開始吃飯。

「湯尼，真對不起啊，佔了你的位置！」在四周高高低低大大小小此起彼落的廣東話中間，如玉用台灣國語向這位廣東帥哥道歉。「沒有關係，我這不是坐得好好的嗎？怎麼樣？今天的花椰菜好不好吃？」湯尼問如玉。「不錯啊，這是廣東味嗎？」如玉問。

「這哪裡是廣東味，把你的電話號碼給我吧，下個禮拜二，我帶你到舊金山的好吃的廣東餐館去嘗嘗真正的廣東味！」果然，下一個禮拜二，湯尼就介紹了如玉到舊金山一家有名的廣東大酒樓，去吃了一頓非常精緻美味的廣東點心。

這才知道湯尼年輕時由廣州暨南大學畢業之後，輾轉到了美國洛杉磯，帶了大批老墨工人們在的廢棄的礦場內種植食用蘑菇，并遇見一位台灣姑娘，兩人婚後生活十分美滿，可惜好景不長，退休後不但愛妻亡故，也沒有子女。

「我的亡妻是一位閩南美食專家，所以我也知道一些舊金山最好的台灣餐館，哪天我也帶你去試試！」所以他們後來又去了好幾家比較好一點的台灣及北方餐館。

「這位廣東帥哥長得還不錯，不過他曾經建議要與我一同去租一套公寓來住，大概他的手頭比較拮据罷？！」如玉與媳婦談天的時候這樣說。

「奶奶，妳在什麼地方遇見這位先生的呢？」媳婦問道。「在舊金山安老自助處成龍康樂中心吃午餐的時候遇見的。」如玉回答。

「那裡的午餐多少錢一頓呢？」媳婦追問。「每人每頓樂捐美金兩元。」奶奶回答道。

「這就看妳啦，妳是寧願遇見可愛善良的老帥哥呢，還是要遇見多金的闊佬呢？要若是找後者，可能就得到那種每頓廿美元，甚至兩百美元的地方去消費才是。」美國媳婦給中國婆婆出了一個好主意。

乖乖龍的東，這麼聰明！難怪兒子對她言聽計從。

A Lucky Day at the Library

Edited by Descartes Li

The poster says that the Main Library has recently launched a weight loss program, and it will be preceded by a half an hour movie in the basement. Seeing this sign, Ruth notices that her vision isn't as clear as it used to be. Trying to find the entrance to the basement, she raises her head and squints her eyes. And, misses a step and stumbles to the ground. With some embarrassment, she finds that is not able to get up by herself.

Fortunately, in America, everyone is on the lookout to do a good deed. So, as Ruth falls, there are several men who come rushing to her rescue. One particularly handsome man steps forward and says firmly to the other would-be good Samaritans: "Don't worry, I'll take good care of her, you can leave us."

Wow! He is so tall and has such good posture! What luck to get picked up by such a healthy and handsome guy. Today must be an extra lucky day! The fall was worth it! Ruth wants to laugh out loud with joy and raise her arms in triumph. What she does instead is to raise her arms so he can help pick her up.

She stands up but is a little shaky, so the handsome man then carries her to the shade of a big tree outside of the library. He grins at her and says: "Look at me, I am almost 70 years old, but I'm strong enough to lift a solid lady like you!" After that, probably wanting to prove his strength, he picks her up once more, this time lifting her feet completely off the ground. He says: "If you exercise with me, maybe you can lose a few pounds."

Having been lifted up three times in a row, Ruth is very pleased. She heard that smiling will tighten your facial muscles, making your face less droopy and therefore appear much younger than you really are. Thinking of this, Ruth quickly nods, trying to smile and lift up her face.

The handsome man sees that she seems to be in agreement with him, so he also smiles and nods. He takes Ruth's hand and says, "Let's go, I will take you jogging and build up your strength!"

He pulls her along by her hand, and they walk through the streets of San Francisco. Eventually, after what seems like hours to Ruth, her entire body starts to ache, and her eyesight is now even worse. The whole world seems not only unclear, but also very distorted. Ruth recalls how farmers would drag the dead carcasses of pigs through the streets of her hometown in China. However, although her legs start to feel as if they are about to fall off, she refuses to admit to herself that she is too old. She didn't emigrate to America to be weak! She grits her teeth, ignores the pain, and reluctantly continues to stumble forward.

More time passes. After a few more blocks, all her blood is now in her head, and her heart is beating so hard in her chest she feels like she is in an earthquake. And while she may be on the verge of death, Ruth struggles to control her desperate breathing. She mustn't show that she is old. But just as she may have to surrender to the demands of her lungs, it seems that the final stroke of luck is hers! A San Francisco cable car clanks to a stop by their side of the street.

"Hey, how about that we take a ride?" Because the handsome man strongly grips Ruth's right hand, she has to use her left hand to point tremulously at the cable car. Her utterance helps to cover her sigh of relief as well as mask a deep inspiration.

"This is for tourists, we need to exercise." The handsome man replies.

Spying a coffee shop from the corner of her eye, Ruth says, "Well, I am a little bit thirsty, let's go in and drink something?"

"Nothing is as healthy as mineral water! Drink this!" he says, and takes out a bottle of mineral water from his backpack. While Ruth uses her left hand to take the water bottle, she also takes the opportunity to breathe deeply two more times.

With this respite, she realizes that the when she fell, her glasses are crooked. No wonder the world is so distorted and unclear! "Looks like my glasses frame is broken, do you know any store to repair glasses?" Ruth asks.

It seems that her luck isn't so bad after all, and the handsome man takes her to a Chinese optician's store. While the clerk is repairing her glasses, Ruth sits in a chair, and, by secretly breathing deeply, is able to catch her breath. She senses that there is better luck to come. Waves of food smells float through the air, and it turns out that next door is a Chinese restaurant. Ruth has an idea.

"Hey, handsome, in order to thank you for treating me so well, I will take you to next door for Chinese goodies!" Ruth says with a coy smile.

He responds heartily, "Have I got something for you! This will keep you healthy and strong. The fat will melt from your body." He pulls out from his backpack two shimmering objects. The look of them renders Ruth speechless for a long while. They are two shiny protein bars for athletes.

Now it's time for them to go their separate ways. The handsome man takes out a fancy business card from his backpack. Pointing at the phone number on it, he says: "If you exercise more than three hours a day, you will be in much better shape and lose a lot of weight. With better tone, you could be very attractive."

He asks her for her phone number. She decides to pretend to be hard of hearing, and bends down to untie, then tie, her shoelaces. She walks away with a smile, a wave, and a feeling that she is very lucky to go home now.

圖書館的艷遇

　　海報說總圖書館最近推出一個減肥的節目，要大家先到總圖書館的地下室去先觀看半小時的電影。如玉最近老是覺得眼睛有點昏花，所以一進圖書館就抬起頭來，眯著眼睛專心一致地尋找地下室的入口。

　　不意腳下踩了一個空，整個人面朝下摔在地上，爬不起來了。

　　美國禮儀之邦，人人以日行一善為榮，所以，如玉這麼一跌，就有數位男士過來搶著救美，其中一位帥哥彎下腰來，雙手將如玉由地上抱起來，口中大喊：「大家放心，我會好好照顧她的，你們走開吧！」

　　哇！高高的個子，挺拔的身量！竟然會遇見這麼善良且健康的帥哥，如玉心中不免竊喜，今天運氣實在不錯，這一跤跌得真是值得呢！

　　帥哥把她帶到圖書館外的大樹下面，重新又把她抱起來一次，放下來之後，對她點點頭，說道：「你看我，雖然已經快７０歲了，可是身體這麼好，像妳這麼胖的女人，我都可以一把抱得起來！」說完之後，大概要表示自己真的是大力士吧，再把她雙腳離地抱起來一次，口中說道：「只要跟我一樣的鍛鍊，保險會跟我一樣健康和強壯！」

　　如玉被他一連抱了三次，芳心大悅，想起大家都說微笑了以後臉上的肌肉比較不那麼下垂，看起來會比較年輕，想到此處，如玉連忙點頭，努力展開一個笑容。

　　帥哥看見她不但沒有反對，而且笑著點頭，增加了他的信心，就牽起如玉的玉手，說道：「走，我帶你出去快走，鍛鍊身體！」

　　如玉腳不點地，被他拖著在三藩市街邊的人行道上向前奔走了好一陣子，不但覺得身體萬分疲累，甚至覺得自己雙眼更加不行了，怎麼整個世界看起來不但格外不清楚，而且還十分歪曲變形了呢？加以雙腿也開始酸痛，但是，既然這位帥哥這麼年輕，自己總不能承認自己太老罷，只好咬著牙強忍疼痛，勉強繼續跟著他向前飛奔。

　　又不知過了多久，雖然覺得血液已經衝到頭頂，一顆心在胸腔裡跳得驚天動地，眼見就要不行了，為了表示自己還不是那麼老，如玉還是不敢大聲喘氣，總算運氣不錯，一輛三藩市叮叮噹噹的纜車恰好停在他們身旁的街邊。

　　「這位帥哥，我們坐車吧！」如玉右手被帥哥牢牢地牽著，只好用左手指那輛三藩市的纜車，借著開口講話的機會，大大地喘了一口氣。

　　「這是給觀光客坐的，不能達到鍛鍊身體的目的。」帥哥反對。

　　「我有點口渴，咱們進去喝點什麼吧？」一眼看見旁邊有一家咖啡店，如玉突然靈機一動，如此建議。

　　「什麼飲料都不如礦泉水健康！那，喝吧！」帥哥一面說，一面由背袋中取出一瓶礦泉水。如玉用左手接過水瓶，借喝水的機會，偷偷地喘了兩大口氣。

　　這才發現，剛才圖書館裡這一跤跌得不輕，眼鏡架都被撞歪了，難怪世界起來模糊而歪曲變形。

　　「帥哥呀，我的眼鏡架跌壞了，你知道那裡有眼鏡店替人家修理眼鏡的嗎？」如玉連忙問道。

　　運氣還算不錯，帥哥果然帶她到了一家中國眼鏡店，　如玉趁店員修理她的眼鏡的時候，坐在顧客的椅子里，偷偷地大口喘著氣，那知接下來還有運氣更好的事，一陣陣食物的香氣由空氣中傳了過來，原來隔壁就是一家中國餐館，如玉突然覚得飢腸轆轆，非得吃些什麼不可了。

　　「嗨，帥哥，為了感謝你對我這麼好，我請你到隔壁去吃些中國美食如何！」如玉笑嘻嘻地問。

　　「這個，我告訴妳，千萬不要亂吃凡塵的食物，看，妳若吃了我的這個，保險又健康又強壯，全身一塊贅肉都沒有！」他由他的背包里掏出兩塊什麼，如玉定睛一看，真是哭笑不得，半天講不出一句話來。原來是給運動員吃的包裝精美的蛋白棒。

　　臨走，那位帥哥向如玉要她的電話號碼，她假裝耳背沒有聽見，彎下腰來解開自己的鞋帶又重新繫上。

　　最後，帥哥由自己的背包里抽出一張印刷得非常精美的名片，指著上面的電話號碼，对如玉說:「如果妳每天再多運動三小時，就不會這麼胖了，其實妳長得還是不錯的，有空打電話給我吧！我是有美國藉的北歐克西米人。」

　　不錯，與帥哥分別之後，如玉覚得：能夠獨自自由自在地回家，真幸運!

原載美國都市報 2017-07-15

Encouraging Her Husband to Seek Adventures

- The story of Yerba Mate tea and grilled steak churrasco

Edited by Adela Karliner (2020-06-22)

We were very energetic when we were young, and enjoyed socializing with our friends. Once, to welcome a new South American researcher, Dr. Diego Garcia, from Argentina, we invited everybody who was working in my husband Luke's laboratory to our house for a dinner party. In Argentina, Diego was an ophthalmologist. He had to work temporarily as a research scholar in my husband's lab because he did not have a license to practice medicine in the United States.

Guests arrived one after another. The primary guest had not yet appeared, so we sat down casually.

"Mrs. Li, I tell you, Dr. Diego Garcia is such a handsome man that he was liked by Judy as soon as she saw him!" my husband's young research assistant, Linda, brought her plate with some food, and sat next to me, told me with a smile.   Dr. Judy Koepnick was a post-doctoral researcher in his lab, smart, beautiful, and very helpful to those who needed her.

"Hey, young one, don't talk nonsense. When I applied for Dr. Diego Garcia to enter the United States, his data listed that he not only has a wife but also has a pair of four-year-old twins, a son and a daughter, Judy is nice to him because this was his first time in the United States, Judy was helping him adjust to the new environment." My husband heard our gossip, so he hurried over to defend his new employee.

While we were discussing them, the two appeared. And indeed, Diego was tall and handsome. We were used to seeing Judy wearing a white laboratory coat. Today, she had changed into a fancy dress and a pair of half-high heeled walking shoes. She looked very different and very refreshing. It turned out that Diego didn't have a car. Judy drove him around to find a place to live and then took him to the post office to collect the parcels from his hometown. That day, Dr. Diego Garcia brought a new packet of Yerba Mate tea from his hometown and a straw-in-a-pipe Yerba Mate cup, which he gave us as a gift.

After the meal, the newcomer Dr. Garcia demonstrated how to pour the hot water into the cow's hoof cup to brew the holly leaves, and then with a silver metal straw to suck the tea. The length of the straw they used was just like the Taiwanese plastic straw we used to drink our boba milk tea, but the bottom of the Yerba Mate tea's straw had a small hole in the filter teaspoon, and yerba Mate tea uses no sugar, so when you suck the mate tea it tastes a little bitter.

"Yerba Mate tea can refresh you, put you into sleep, help with weight loss, lower blood pressure, and also reduce the lipids in your blood!" He told us in English with a cute Argentinan accent. Everyone was very interested.

"That's great! Reducing blood lipids is perfect for me! I'm trying to lose weight." Judy cried with delight.

"It's easy. I will ask my wife to send you another one, in our hometown, 'handing a Yerba Mate teacup waved through the city' is a familiar scene!" Diego said. I thought when he called his wife at home, and he would tell his wife how sweet his new friend Dr. Judy Koepnick was to him.

Soon, Judy invited her boss Luke and the boss's wife, me, to her home to eat her newly-learned Argentinan dish churrasco, and it turned out that Diego had moved into Judy's house to live with her.

Steak is very delicious, tender, and fragrant with a distinctive smoky flavor. It is so tasty that you would even want to swallow your tongue! Judy told us that she had the original recipe from Mrs. Garcia in Argentina.

"How about MY Argentina's home cooking? Can it reduce some of your nostalgia?" Judy tenderly whispered in Diego's ear and asked him with a sweet smile.

"American beef is processed after a long time aging, much more refined than our native Argentinan beef!" Diego looked at Judy with emotion, nodding gravely. In this world, men were born not to resist the beautiful women's tender favor!

Among the world's beef, United States beef is top quality. To pursue the best, Americans use scientific methods to raise first-rate cattle to start with, and they don't sell the meat immediately after slaughter, but hang the beef (or package) in the prescribed temperature and conditions to make the meat better. This process takes more or less 28 days, so the best meat sometimes had been slaughtered a month ago!

Then they talked about how Argentina's government took

Diego's wife as a hostage, forbidding her to leave the country. In order not to be deported by the United States Immigration Service, Diego intended to marry Judy, an American citizen, and Judy would come forward to adopt Diego's son and daughter so that they could attend to high school here have a better future.

Later, after we retired and moved to Florida when Diego get a green card, obtained a New Jersey eye doctor's license and opened an ophthalmology clinic; Judy quit her job as a post-doc in the lab, using her local business knowledge and connections to help her husband Dr. Garcia's practice. Judy's family had owned a local grocery store started by her grandmother. She is the third generation European-American.

To celebrate the twin's graduation from a prestigious American university, their father and stepmother held a large party for hundreds of people. We had long retired, we sat on the riverside balcony of our Florida house to see the videotape they sent. As I sipped my yerba mate tea, I broke out in big sweats. The tea I sucked in my mouth became extremely bitter and hard for me to swallow because my thought went to the twins' birth mother, Dr. Garcia's ex-wife, who was held hostage in Argentina. How did she feel when she was watching this tape, was her heart comforted that her motherly sacrifices had finally resulted in the twins' graduation from a famous American college?

Or, she felt as the famous Chinese poet Wang Changling's young woman in his poem 'Complaining woman,' "Suddenly seeing the color of Willow on the river bank, She repented that she had encouraged her husband to seek adventures away from home"?

What do you think?

悔教夫婿覓封侯

文／佘國英

——瑪黛茶 yerba mate 及 churrasco 烤牛排的故事

我們年輕時精力比較旺盛，很喜歡與朋友們交往。有一次，為了歡迎一位新由南美來的研究學者迪亞哥 加西亞 Diego Garcia 博士，特地請了全實驗室的人來我家作客。迪亞哥在巴拉圭本來是眼科醫師，因在美國尚未有執照，只好暫時在我老公的實驗室裡做一名研究學者。

陪客們一個一個進來，主客尚未出現，大家就隨意坐了下來。

「師母，我告訴妳，這位加西亞醫師真是一位大帥哥，一來就被茱蒂看上了！」我先生年輕的女助琳達一面吃東西，一面笑嘻嘻地告訴我。

「呔，小孩子，不要胡說，當初我申請迪亞哥醫師入美國境的時候，他的表格上明明白白的填了不但有老婆，還有雙胞胎一兒一女呢，茱蒂是因為他初來乍到，人地生疏，才幫助他熟悉一下環境。」我老公聽見了，連忙走過來替他們分辯。柯博士茱蒂 Dr.JudyKarliner 是他實驗室內的後博士女研究員，長得聰明、漂亮、非常樂於助人。

正在討論他們，兩人就出現了，果然迪亞哥又高又帥，十分風流倜儻，平常看慣了穿白色實驗制服的茱蒂，今天換了一件花色洋裝，腳上登了一雙半高跟的步行鞋，也令人耳目一新。原來迪西亞沒有車，茱蒂先開車帶他去找房子，後來又帶他到郵局去領取家鄉寄來的包裹，所以耽誤了一些時間。那天，加醫師帶來了他太太由家鄉新寄來的一包瑪黛茶和插了吸管的瑪黛茶杯，送給我們當作禮物。

人手一杯，招搖過市的瑪黛茶

飯後，加醫師就示範如何用熱水沖泡進牛蹄杯中的冬青葉乾 Ilex，然後用銀色金屬吸管吸茶，吸管與我們台灣人吸波霸奶茶一般長短，不過瑪黛茶吸管底有小孔的濾茶渣匙，而且瑪黛茶不放糖吸進嘴裡有些苦澀。

「瑪黛茶可以提神又可以安眠，還可以通便減肥，更可以降血壓，減血脂喲！」他告訴我們，大家都非常有興趣。

「太好了！減血脂！我正想減肥哩。」茱蒂很高興地嚷道。

「這太容易了，我要我太太再寄一支來給妳好了，在我們家鄉人手一支瑪黛茶杯招搖過市呢！」他對柯茱蒂博士說，我想，他打電話回家時，一定會告訴他太太關於在美國實驗室內的新朋友柯茱蒂的如何熱心幫助他的情形。

不久，茱蒂邀請她的老闆盧馼及老闆娘我到她的家中，去吃她新學會的烏拉圭家鄉名菜 churrasco 香木烤牛排，原來迪亞哥已經搬進茱蒂的房屋与她同住了。

churrasco 香木烤牛排

牛排非常鮮美適口，加上有一种特別烤肉的香味，又嫩又香，入口即化，真叫人舌頭都要吃掉了！茱蒂告訴我們是她向遠方的加太太討來的祕方。

「我做的烏拉圭家鄉菜怎麼樣？能夠聊解一些你的鄉愁嗎？」茱蒂低聲很溫柔地笑著問迪亞哥。

「美國的牛肉是經過，花長時間冷藏處理 age 過的，比我們土產烏拉圭牛肉肥潤鮮美得多了！」迪亞哥很動情地一面看著茱蒂，一面深深地點著頭，男人生於世上，最難消受的就是美人恩喲！世界上的牛肉，以美國的牛肉最為鮮嫩，因為美國人養牛，用科學辦法飼養，追求牛肉的品質，而牛肉屠宰以後，并不立刻拿出去銷售，而是要將牛肉掛（或包）在規定好的溫度，特定的狀況下，使肉質更好，這种 Process，少則數天，多則 28 天，所以最好的牛肉，有時是一個月以前屠宰的呢！

接著他倆很無奈地談起烏拉圭政府要扣押迪亞哥的太太為人質，不許她出境，而迪亞哥為了不被美國移民局遣返，打算要与美國藉的茱蒂先結婚，并由茱迪出面領養迪亞哥的一兒一女到美國來讀中學，以便孩子們將來有好一些的出路。

後來，我們退休後搬到佛羅里達，有了綠卡的迪亞哥考到新澤西州眼科醫師執照，並開了一家眼科診所，柯博士茱蒂也辭掉在實驗室內的後博士工作，運用她當地人的企業知識，全力協助她的丈夫加西亞醫師的營業，她是當地人，從祖母開雜貨店起，傳到她已經是第三代了！

最近，又收到他們倆人為了慶祝雙胞胎兄妹由名校大學畢業，開了一個數百人的大型的派對的錄影帶，我們退休已久，只想懶坐在河邊自家的陽台上看看他們寄來的錄影帶，雖然裡面的客人一個也不認識。我一面吸著瑪黛茶，一面發出冷汗，吸在口中的瑪黛茶愈發苦澀不能下嚥。因為想到了在烏拉圭被扣作人質雙胞胎的生母，那位迪亞哥的前任太太，她的心裡是安慰呢？還是像

王昌齡《閨怨》中的少婦：《忽見陌頭楊柳色，悔教夫婿覓封侯。》呢？

Edited by Wendy Chou

"Hello, is your name Jade?" One day, when Jade was walking on the street hurriedly, she was stopped by a man who sat in the car with the window open.

"Oh, it's you! You're Little Jasmine's... friend!" In order to be polite, Jade had to stop. She stood at the side of the car, thinking how lucky it was that she had recently cut her hair short, and hoped that her thick hair band might cover the wrinkles on her forehead.

"But, somehow, she's not my friend anymore," he said. As the guy looked somewhat depressed, Jade felt very sympathetic toward him, so she took a closer look at him. She realized that although this person was not very young anymore, his face still looked quite handsome. She thought that he must have been a handsome man when he was young. In fact, he still could be considered a great-looking senior. She even felt that it was a pity that little Jasmine was no longer his friend any more.

"Not your friend anymore? What is your name? " Jade asked him, then squeezed out a smile in a hurry. Smiling pulls the facial muscles, makes people look less tired, so she heard.

"My name is John, where are you going?" He seemed to want to talk to her, and did not particularly care about the expression that Jade just squeezed out.

"Me? I am having writer friends coming from Hong Kong, Taiwan and Canada, and, I am.... " Jade said. She specifically straightened her back, lest this handsome man see her hunched, aging look.

"Is it so? I also write. How about inviting your writer friends to dinner?" he asked.

"Okay, I'll ask them and give you an answer!" She replied hurriedly, eager to walk away, for fear that her mature age would be clearly visible to him.

"Well, please take my phone number, I'll be waiting for your reply." The man was very enthusiastic.

Jade went home and asked her friends. The next day, she gave John a phone call as she had promised, and told him that every one of her friends agreed that they could not allow him to spend his money, so they politely refused his invitation.

"Please don't hang up the phone so fast!" He said hastily; "I hear you can translate, would you like to translate my English book   into Chinese?"

What she did not expect was that through the phone, the guy's masculine voice was extremely low and sexy.

"Yes, I would love to translate a meaningful English book." Jade was also anxious to be his friend.

"Well, how about having coffee with me tomorrow at noon? Let's talk about the details over coffee?" His Southern American accent was really charming.

The next day, as John was sitting in the coffee shop, he saw Jade step into the shop. He stood up very politely to order a cup of coffee for her. When he stood up, he was so tall, so big, and so full of male charm. Jade suddenly felt her heart was beating faster, her soul was flying…. She wanted…..

"Why are you wearing a mask?" John asked her in amazement.

"Because the fog in San Francisco is very heavy, I am afraid of catching cold!" Jade answered uneasily. She could not tell him that she was afraid that he could clearly see the creases on her face and her puffy jawline!

Fortunately, he was busy kissing her hand, and did not continue to investigate anything pertaining to her mask.

After the coffee, John invited her to his house to enjoy a hot summer barbecue the next day.

Of course Jade certainly went.

"It's very hot today, give me your scarf!" John's hands reached out for her scarf.

"No need, my scarf and clothes are matching!" Jade refused to untie her scarf. At least the scarf could cover the wrinkles on her neck.

Fortunately, he was kissing her lips attentively. Both their bodies and souls were generating so much heat that wrapping or unwrapping   the scarf was entirely not important anymore.

The Chinese have a saying; No matter how ugly a woman is she has to face her in-laws! When you are dating a man you cannot always use your hair band to hide your wrinkles on your forehead, forever wear a mask on your face to cover the creases near your nose, nor can you permanently wrap a scarf around your neck! What can one do?

Fortunately, luck was with her. Jade found an advertising notice on Facebook.

You are invited to attend the Silhouette instaLift Lecture held by Dr. Victor Liu's clinic

Learn how to revitalize your face and give yourself another chance to be young again.

Eyebrow-pulling will solve the problem of sagging of the eyes, cheeks, facial line, lip line, the face contour, the double chin, and the neck line.

Drinks and snacks available on site

Date: August 21 (Wednesday)

Time: 7 p.m.

Where: 1720 El Camino Real Suite 200, Burlingame, CA 94010

Spots are limited, please register with us immediately, tel: (650) 697-8888

Jade invited her close friend Bin Mei Mei to listen to the lecture with her, and Dr. Liu's seminar was a great success, captivating the audience.

Knowing that everyone has the chance to be young again, every listener became hopeful.

"Dr. Liu, do you think I could be saved?" Jade was eager to raise her hand, and spoke enthusiastically.

"It's just that one needs to be young to go through the Silhouette instaLift." Dr. Liu replied honestly.

"Is it too late for me?" Jade's heart sank.

"No worry, don't you worry, you can still go through the mini plastic surgery; it certainly will let you return to youth, and you will be young and beautiful as before!" Dr. Liu reassured Jade.

"Really? Is it true?

Jade was very excited!

Jade did all the CMC, CMF EKG...., and passed all these with flying colors, underwent the surgery and waited through a brief recovery period.

Three weeks later...

He came and waited for her downstairs at her place. Jade had permed her hair, revealing her new smooth, wrinkle-free forehead.

"It's getting cold. Where are your mask and scarf?" he smiled and asked Jade, noticing her beauty, as she got into his car.

"Didn't you know the whole earth is warming up? Sir! It is the effect of Global Warming! " Full of self-confidence, Jade replied to him in English with a big smile on her face.

"I'm so lucky to meet you in this life, you are a beautiful young lady," John praised sincerely.

Surgeons think that plastic surgery can improve the value of one's look. Fortune-tellers are convinced that plastic surgery can even upgrade your karma; We sincerely wish that this handsome senior couple will remain forever young and eternally in love!

整容可以提昇命運

　　「哈嘍，妳的名字叫如玉罷？」這天，如玉在街上匆匆行走，被一位開了車窗，坐在車中的男人叫住。

　　「哦，是你！你是小茉莉的… 朋友！」為了禮貌起見，如玉只得站在車邊回話，好在最近剪了一個妹妹頭，希望瀏海能夠遮住自己額頭上的皺紋。

　　「不過，不知為什麼,她已經不 是我朋友了。」這人很沮喪地說道，如玉見他訴苦的樣子很可憐，再仔細一看，只覺得這人雖然不年輕了，但是面孔五官長得不失英俊，想當初一定是帥哥一名，其實，到目前還可以算作資深帥哥，她甚至覺得，老大不小的小茉莉不跟他在一起，實在十二萬分可惜。

　　「是嗎？你的名字是…？」如玉問，急忙之中，擠出一個笑臉，聽說，笑容扯動面部肌肉，使人看起來比較不那麼蒼老疲倦。

　　「我叫約翰，妳到那去？」他似乎很想與她講話，并沒有特別在意如玉硬擠出來的表情。

　　「我?…, 我有作家朋友們由香港、台灣及加拿大來，我…。」如玉說，特地挺了挺胸，免得被這帥哥看出自己彎腰駝背衰老的樣子。

　　「是嗎？我也寫作，我請妳的作家朋友們吃飯如何？」他問道。

　　「那我先問他們一下再回答你罷！」她匆匆地回答, 急欲走開,生怕自己成熟的年齡被他看得一清二楚。

　　「好，請妳把我的電話號碼拿去，我專心等妳的回音。」這人很热心地囑咐。

　　如玉回去後問了她的文友們，第二天，依約給了他一個電話，告訴他說,大家一致認為不能隨便讓人破費,客氣地回絕了他。

　　「請妳不要這麼快掛斷電話!」他急急忙忙地說:「聽說妳會翻譯？妳願意把我的英文書,The turning point of my life -- "Powerhouse Road" 翻譯成中文書嗎？」沒有想到，電話裡傳過來這傢伙的男性聲音，竟然十二分低沈性感呢。

　　「是喔,我也很想翻譯一本有意義的英文書,我一生的轉捩點 -- "能源廠之路"，好。」如玉也很想做他的朋友。

　　「那好，我請你明天中午喝咖啡, 咱們在咖啡廳談一談細節罷。」他那美國南方口音,著實迷人。

　　第二天, 坐在咖啡店的約翰看見如玉進店, 很禮貌地站起身來替如玉要了一杯咖啡。他這一站，顯出他高高魁偉的身材，充滿了男性的魅力。如玉突然覺得臉紅心跳，魂飛天外。恨不得⋯⋯。

「妳怎麼戴了一個口罩？⋯⋯。」約翰很詫異地問她。

「因為最近舊金山的霧霾很重，怕傳染傷風！」如玉不安地回答。總不能告訴他是怕他看清楚自己臉上的法令紋和木偶紋罷！好在他正在忙著親吻她的玉手，沒有再繼續追究口罩的事情。

喝完咖啡，約翰約如玉次日到他家去吃炎夏燒烤。

如玉當然如約去了。

「今天天氣很熱，把妳的圍巾給我吧！」約翰伸出手來等着接如玉的圍巾。

「必要喔 我這圍巾和衣服是配套的！」如遇拒絕解開圍巾，至少,圍巾可以遮掩頸紋喲。

幸好他正在專心地吻著她的嘴唇，兩個人的身心全部都發出驚人的熱氣，圍不圍圍巾已經很不重要了。

醜媳婦總要見公婆的，怎麼辦呢？她總不能天天掛著瀏海、帶著口罩、圍著圍巾見面呀！幸好，人算不如天算，如玉在面書上發現一個廣告通知。

誠摯邀請您參加廖敬仁醫生診所舉辦的鈴鐺線拉提講座

由廖醫生向你講解鈴鐺線拉提 (Silhouette InstaLift)

可以怎樣回復青亮麗面容，給自己再年輕一次的「肌」會。

鈴鐺線可以眉型拉提，對眼尾下垂、蘋果肌下垂、法令紋、木偶紋、嘴邊肉、臉部輪廓、雙下巴、頸部線條的問題，有顯著的效果。

現場提供飲品及小吃

日期： 8 月 21 日(星期三)

時間：下午七時

講座地点: 1720 El Camino Real Suite 200, Burlingame, CA 94010

名額有限，請立即向本處報名留位, 電話: (650) 697-8888

如玉就約了好友彬小妹一同去聽演講，只見廖醫生講得頭頭是道，舉座動容。知道每人都有再重新年輕的機會，每位在座的聽眾們也都個個心花怒放。

「廖醫師，你看我還有沒有救？」如玉熱心舉手，踴躍發言。

「你嗎？鈴鐺拉提要趁年輕，所以...。」

「太晚了嗎？」如玉的一顆心掉在谷底。

「不急，不用急，你還可以做面部開刀拉提的整容手術，一定讓妳恢復青春，美麗如昔！」廖醫生很慈祥的安慰如玉。

「真的?真的嗎？是真的可以嗎？」如玉太高興啦！

做了 CMC， CMF EKG....， 過五關，斬六將，終於經過了整容手術，恢復期三個禮拜，只需要三個禮拜。

三週過去之後...。他開車來了，坐在如玉家的樓下等她， 只見她燙了新頭髮，露出她嶄新光滑沒有皺紋的額頭。

「天氣漸漸冷了，怎麼反而看不見你戴口罩、圍圍巾呢？」他由車內一眼看見美容過的如玉，笑嘻嘻地問她。

「君不見整個地球都暖起來了嗎？The effect of Global warming 呀!」充滿了自信的如玉笑嘻嘻地用英文回答他。

「我真幸運，今生能遇見妳這位年輕貌美的佳人。」約翰誠懇地說。

外科醫師認為整容可以改進顏值，風水師確信整容甚至進而提昇命運，咱們在此謹祝本文中的資深帥哥以及大齡熟女:青春不老，爱河永浴！

The 21-century modern matchmaker

Edited by Adela Karliner (08/26/2020)

While Jade was home attentively tapping on her computer's keyboard， her cell phone suddenly rang.

"Jade， why are you still home? Isn't today the day when my hometown girl August Moon and the gentleman Clay， whom she met through a matchmaking website， are seeing each other in person? How come you， the lady who promised to work as a translator， are still at home?" On the mobile phone came the nervous voice of the newly divorced Honey Flower.

"Is it today? If you didn't call， I would have forgotten!" Jade was startled; her brain was still not very clear.

"Hurry! August Moon is already there. You had better be fast! Remember， Jade each of them will pin a big red rose on their chest as a sign of identification， so they are easy to find." Honey Flower hung up the phone in a hurry.

Since her friends trusted her， she had better be loyal to her friends. Their meeting place is a Chinese Restaurant in Chinatown， Oakland. Luckily， Jade lives across the street from this very Chinese restaurant.

Jade hurriedly turned off her computer， grabbed her purse， caught the elevator， and got out of their apartment building in a jiffy.

While she was standing on the street and waiting for the red traffic light to change color， she suddenly saw a reflection， her face reflected in the glass window of a silver SUV. The car's engine was still running.

Oops， Jade realized that she had forgotten to put on her makeup!

Jade could not help but take out a lipstick from her purse. Against the glass window， she applied it to her lips. While combing through her hair with her fingers， she heard the deafening sound of car horns blasting from all the directions.

Suddenly， Jade found the mirror in front of her rolled down. She had to bend down lower her body to look in the mirror. Before long， Jade realized that the mirror had disappeared! She found that the driver of the SUV had pushed his remote control. He had rolled his car window down.

"Hello, lady, have you done with your makeup? I think you are beautiful enough without the makeup! The traffic light has turned green. All the cars behind me were impatient, desperately honking their horns, urging me to move ahead!" The old gentleman said apologetically. He had a pair of kind brown eyes and no hair on the top of his head. He seemed to be a sweet-natured man.

Jade was startled. Then she suddenly realized that she was checking herself in the middle of heavy traffic and applying her makeup in front of his car window! She was even more embarrassed by what he just said.

Regardless of the color of the current traffic light, Jade immediately fled to the sidewalk with a blushing red face, trying to merge within the crowds.

As soon as Jade stepped into the restaurant, she saw a beautiful-looking, young Chinese woman wearing a large red rose on her chest and sitting nervously in her seat.

Jade walked toward her, then sat next to the woman.

"Here he comes, he's coming!" Before Jade had time to say hello to August Moon, she heard August Moon murmured excitedly.

Can you guess who August Moon's date was? It turned out to be the old gentleman who had just rolled the window down on Jade!

He also wore a big red rose on his chest plus a teasing smile on his face.

Fortunately, August Moon was very serious. She asked Mr. Clay directly: What is his monthly income? Does he own any real estate? She wanted to get married legally and formally. If she is married to a citizen, she will be able to stay in the United States and receive half of his Social Security and pension while he is alive. After his death, she will be able to remain in the United States and receive all of his benefits and pensions.

Wow, Jade was surprised that August Moon might not be able to speak English, she knew all these details very clearly.

"Tell him that I am not a greedy woman, he can leave all his disposable cash and real estate to whomever he wishes, my purpose is that he shouldn't waste any of the resources stipulated by the law and government." Jade translated August Moon's Chinese into English immediately.

"I need a woman to be with me, to do the housework, and it doesn't matter what kind of food she cooks, because I also love Chinese food, " the American gentleman Mr. Clay was also straightforward. Therefore, her translation job was almost effortless.

"Now, that we have reached an agreement. Let's all go to my house and take a look at my living environment, " Clay proposed. August Moon agreed right away.

Because both sides were frank and straightforward, Jade, the matchmaking translator, also thought that she had accomplished her mission, so she intended to go back immediately to continue knocking on her keyboard.

"Please, Jade, please don't go back home yet, we will need you to do the translation for us after we arrive at my home!" Mr. Clay asked their matchmaker.

Mr. Clay drove his silver SUV, and brought his date August Moon and their translator, Jade, to his private home in a beautiful walnut grove in the Napa Valley. His white house has two bedrooms and two separate living rooms.

"It would be great if I planted a vegetable garden here and had a flock of chickens and ducks." Autumn Moon was from a rural Chinese village. She was delighted with her date; "the minute I get my citizenship, I will apply for my son to live with us."

Suddenly, Clay unexpectedly pulls Jade aside and asks her if she would take care of his home. The reason is that he likes Jade more.

"The BART doesn't go through here, but I can buy a used car for you as a means of transportation, " the old gentleman is quite enthusiastic!

"But I have my own house. I am also receiving my Social Security and pension. I can buy a used car to drive if I want to. Why should I take care of your home?" Jade answered without any hesitation.

"Hey, didn't you take good care of your late husband?" Mr. Clay asked.

"That's different. We went to college together. My late husband took good care of me all my life. We had 54 good years together." Jade told Mr. Clay honestly， "August Moon is prettier and younger than me. She will be able to take good care of you until your last day!" Jade pointed out the situation very frankly.

"But I can't communicate with her!" The old gentleman hesitated.

"You can buy an iPad or iPhone for her， then you two can use the 21st century's modern matchmaker to communicate，" Jade had found the old gentleman a great solution to his problem.

Jade was very proud of herself.

廿一世紀的新式紅娘

　　如玉正在家中專心地敲電腦的鍵盤寫作，她的手機突然響了起來。

　　「如玉，你怎麼還在家中啊？今天不是我家鄉同村姑娘秋月與網站上透過經紀人認識的男朋友克萊相親的日子嗎？你這個做翻譯的紅娘怎麼還在家裡呢？」手機中傳來新近離婚閨蜜翠花緊張的聲音。

　　「是今天嗎？你不打電話來我都忘了！」如玉嚇了一跳，腦子還有點轉不過來。

　　「快快快，秋月早就到了，妳快快去吧！記住，男女雙方都各帶一朵大紅玫瑰花做辨認的標誌，所以很容易找的。」翠花急急忙忙掛斷了電話。

　　既然受人之託，就該忠人之事，好在相親的地點就在奧克蘭中國城內她住處對街的一家中國餐館。

　　如玉匆匆關上電腦，隨手抓起皮包，急急趕進電梯下樓，飛奔出了自家公寓大樓，站在街邊等那紅綠交通燈轉換顏色的時候，恰好在她面前停了一輛銀色的 SUV 大車，她就朝大車的玻璃窗子瞄了一下，不意看見玻璃上映出自己的影子，哇噻，竟然忘記了整理容顏！

　　如玉忍不住由皮包中掏出口紅來，對著玻璃窗中映出來的素臉，塗了一下口紅。正在用手指梳理一下亂髮的時候，聽見四周街上的汽車喇叭聲突然此起彼落，振耳欲聾地響了起來。

　　突然，如玉發現她面前的鏡子一直向下退，只得把身子低下來就著鏡子，不料才彎下腰，鏡子已經退下不見了，仔細再看，原來是玻璃車窗被該車的駕駛搖了下來。

　　「這位美麗的女士，妳畫好妝了嗎？交通燈變成綠色了，後面的車子都不耐煩啦，拼命按喇叭催我們呢！」這位有著褐眼、但是沒有頭髮的老先生，笑嘻嘻、好脾氣地對如玉説。

　　如玉被他點醒，突然猛省自己原來站在街邊對著人家的車窗玻璃化妝，嚇了一跳，顧不得街上的紅綠燈目前是什麼顏色，立刻紅著臉逃到對街的人叢之中。

　　一踏進餐廳，就看見一位眉清目秀、年紀適中的中國女子，胸前戴了一朵大紅色的玫瑰緊張地坐在座位上，如玉快步走過去，坐在那女子的旁邊。

　　「來了，他來了。」如玉還沒有來得及與秋月打招呼，只聽見秋月很激動地說。

你道來的是誰？原來就是剛才那位把車窗搖下來的那位老先生，他的胸前也戴了一朵大紅玫瑰花，臉上帶著揶揄的笑容。

幸好秋月非常認真，急忙開門見山，直接了當地問克萊老先生每月收入多少，有沒有房地產，她想要法律上正式結婚，因為只要結婚她就有可以有留在美國的身分，在男方活著的時候，她可以領他社會福利金的一半，也可以領到法定的家屬的退休金。在他過世以後，她還是可以留在美國，領取他全部的福利金以及退休金。

哇，沒有想到不諳英語的秋月對這些細節居然如此瞭如指掌。

「告訴他，我並不是一個貪心的女人，他的動產現金和不動產房屋，他要留給誰就是誰，我的目的只要不浪費公家規定的資源就行。」如玉連忙照實把秋月的中文翻譯成英文。

「我需要一個女人作伴，整理家務，只會做中國飯也沒有關係。」美國老先生克萊也十二分乾脆，如玉這翻譯當得也很容易。

「既然我們對雙方的條件全部達到共識，那就到我家去看看我的房子和居住環境吧。」克萊提議，秋月立刻同意。

因為雙方都很直接了當，所以做翻譯的紅娘也以為很容易地就己經達成了任務，打算立刻回去繼續敲她的鍵盤。

「如玉女士，你先不要回自己的家，我們還需要你來替我們做翻譯吶！」克萊先生向如玉請求。

克萊老先生開了他銀色的ＳＵＶ，帶了相親对象秋月及翻譯紅娘如玉一同到他在納帕谷邊山上座落在美麗核桃樹林中一個很樸實房屋的獨立家園，白色的房子裡有兩間臥室，兩個客廳。

「這裡如果開個菜園，養一群雞鴨，那就更好了。」由中國農村來的秋月十二分滿意這次的相親對象。

這位老公公竟然出乎意料地把如玉扯到一邊，問她願不願意照顧他的家？理由是他比較更中意如玉。

「我這裡雖然沒有捷運，但是我可以替妳再買一輛二手汽車給妳作為交通工具。」老先生還挺熱心的呢！

「可是，我有自己的房子住，有自己的社會福利金以及退休金，可以自己去買一輛二手車來給自己開，我為什麼要照顧你的家呢？」如玉雖然還是很客氣，但毫不猶豫地回答。

「嘿，妳不是照顧了妳過世的先生一輩子嗎？」美國老先生克萊問。

「那不同，我們是大學同學，我十九歲就認識了我先生，他照顧了我一輩子，我跟他有54年的交情，我替他生孩子、煮飯、整理房子是心甘情願的！」如玉把自己的真實想法告訴克萊老先生。

「秋月女士比我漂亮、又比我年青，她才能真正好好照顧您到最後一天啊!」如玉很誠實地指出來。

「可是我與秋月不能溝通啊!」老先生還是猶豫。

「您去購買一個 iPad 或 iphone 給秋月，兩人以後可以用廿一世紀的新式紅娘 Google 翻譯軟體來溝通，不必要我這個真人紅娘來做翻譯。」如玉非常得意地告訴老先生她解決問題的辦法。

The old fisherman and the new neighbor

Edited by Descartes Li

One day, Old Fisherman Yen ran home, and told his wife excitedly; "Jade, good news! There's a new strong and handsome neighbor next door!"

"What does the handsome guy next door have anything to do with you? " asked Jade.

"Well, at this point, we need a young and strong man to operate the boat, isn't it good news?" Old Fisherman Yen could not hide the fact that he was very much pleased.

Eddie, who was White, was from the Northeast, where he had inherited his grandfather's wholesale business, which had sold flowers to various flower shops in New York City. Unfortunately, his products could not compete against the inexpensive imported South American flowers, which were shipped to all of the city's supermarkets. In addition to his financial troubles, his wife had divorced him and took their daughter away with her. Heading south to Florida, Eddie sold his New York flower business plus his home. Comparing Florida real estate prices to prices in New York was like comparing the earth to the sky! He immediately bought the two-storey home next door to Yen's house and used the rest of his money to buy a floating pontoon boat. Just when he was feeling immeasurably self-satisfied with his new financial arrangement, he found that there was no spare money in his pocket to fill his boat with gasoline.

Mr. Old Fisherman Yen was very happy. "It's just perfect, he is young and strong, I can pay for the gas. We two will work together, and out we go to the Gulf of Mexico to catch big fish!"

But the old fisherman and the new neighbor had barely gone fishing together a handful of times when the old fisherman passed away in the nearby Seven Rivers Hospital.

Since Jade and the old fisherman had been married for 54 years, and were so used to each other, Jade suddenly felt that her life was overwhelmingly incomplete. If it were not for Eddie coming over to bring her food and keep her company, she felt she would not have been able to go on with life.

Jade missed the old fisherman's hundred kinds of consideration and thousand ways of gentleness. Soaking in her own tears,   she said: "I want to sell his fishing boat,   because each time I see it,   it reminds me of my old fisherman." The old fisherman had left behind his fishing boat in the yard in a shed.

Eddie said: "I'll call the agent who works for the boat company next to Route 19 and let him haul the boat to their show room. They will put a "For Sale" sign on the boat. There'll be more potential customers over there. And it won't be in your yard. It would be more convenient to have them sell it for you. Also,   safer. There might be riffraff poking around to look at an old boat in your yard." What an excellent idea! Jade would not have thought of it herself.

Once the boat was hauled away,   Jade asked: "Since there will be no boat,   how about selling the boat shed as well?"

Eddie replied. "The shed is not the same as the boat,   it can't be hauled away. And people will want to come and take a look at it before they buy it. Maybe you should have it simply dismantled and taken away. Let me think about it first."

A few days later,   a rugged local man named Harrison, drove in his pickup truck with his tools. He had responded to Jade's request on Craigslist. "Mr. Harrison,   you have to tear it down completely and leave nothing behind,   because I don't have any tools!" Jade said to him.

Harrison quickly broke down the shed,   but before taking away the pieces,   he asked: "Oh,   ... Don't you have a husband? Are you alone?" Whereupon Jade and Harrison had a lengthy discussion of what had happened and her recent sad circumstances.

Eddie,   after noticing Harrison's truck drive away,   went over to talk to Jade. Eddie said uneasily: "Jade,   I saw Harrison talk to you. It seemed like a long talk. What did he have to say?"

"Ah,   he was just asking about my situation".

"You have to be careful,   Harrison is a man with a reputation around here." Even though Eddie was the new neighbor,   he was quick to learn about the local culture.

The next morning Harrison,   in his big leather boots,

thumped up to main door of Jade's house.

Before Jade realized what was going on, she found that Harrison was sitting in her living room, said: "Hey, Jade, make me a cup of coffee, will you?!" Jade was really startled, and confused by this request. Fortunately, a moment later, her front doorbell rang.

Jade hurriedly opened the door, and found it was Eddie! Feeling much relieved and somewhat grateful, she made two cups of coffee for the guests to drink.

In the following month, Harrison came several times to visit, but each time, Eddie quickly showed up. Eventually, Harrison got the message and stopped coming around.

Soon after that, Jade, while reminiscing about the old fisherman, totaled her car by driving into stand of small trees. It was Eddie who gave Jade rides to the local supermarket to buy food. And in order to thank him for his kindness, Jade often took Eddie out for a lunch afterwards.

One time, Eddie suddenly stopped chewing his sandwich and asked Jade who was sitting across from him. "Hey, Jade, there's a blonde woman sitting over there starring at you. Do you know her? "

"Eddie, why would a woman stare at me? She must be staring at a handsome guy like you!" Jade turned to look, and sure enough, it was a woman she knew.

"Hello, Lisa, you don't have to go to work today?" Jade knew Lisa well, a postal worker at the post office in their small town. A few weeks after the death of the old fisherman, thinking she might move back to China, Jade had reapplied for a Taiwanese passport. A new photograph was needed for the passport, and Lisa was the one who took the photo. Jade had told Lisa about the death of the old fisherman and Lisa and she had developed an acquaintanceship.

"I'm taking an early retirement from the post office next month, and I'm looking for a cheaper place to stay after I leave. I've decided to take full lunch breaks", replied Lisa.

"Jade, isn't your downstairs empty? How about renting it to Lisa?" Eddie suggested.

Jade was a little hesitant. "Rent my downstairs to Lisa? Sure, but I'm going to move to California to be near my son and his family. I'm selling the house，what if my house is sold?"

"Don't worry until it is sold!"

So soon thereafter，Lisa did move in to Jade's downstairs rooms. They were compatible and Jade found much solace in having this new companion. But Jade did sell her house，and moved to California as she had planned. Jade experienced an unexpected sense of loss from leaving Lisa and Eddie.

After settling in California，Eddie emailed Jade:

"Hi Jade

As you predicted，the new homeowner didn't want to rent the downstairs out，so Lisa moved in to my downstairs. A few months ago，my parents came to visit and stayed downstairs，so Lisa moved upstairs，first temporarily and now it seems permanently!

We miss you，but we are glad you are happy out there in California. If you are ever back in Florida，don't hesitate to visit Lisa and me. You can stay downstairs!

Love，

Eddie"

隔壁的護花使者

一天，老漁翁先生興致勃勃地跑來告訴如玉：「老婆，告訴你一個好消息！我們隔壁來了一個年輕力壯的大帥哥！」

「來了帥哥跟你有關係嗎？」如玉反問道。

「這一下好了，我們有了年輕力壯開船的生力軍，這還不是好消息嗎？」老漁翁先生掩不住高興。

這位帥哥老美艾迪先生原來在紐約市繼承了祖父的鮮花批發企業，為紐約市內各個鮮花店做批發鮮花生意，可惜敵不住由南美各地空運到超市的鮮花攤上的價廉物美，在強烈的新舊競爭中敗下陣來，加以太太帶著她自己的女兒下堂求去，艾迪哥只得把紐約的生意和房屋急急賣掉，南下到佛羅里達來另求生路，發現此地房地產價錢與紐約相比，真有天壤之別，於是就立刻將李家隔壁水邊的兩層樓房買了下來，又將剩餘的錢買了一隻浮筏 pontoon boat，正在沾沾自喜，自以為得計之時，這才發現，佛羅里達的汽油雖然便宜，但是他的口袋中已經沒有餘錢來替他的船隻加滿足夠的汽油了。

「這不正好，他年輕力壯，我可以付汽油的錢，兩個人通力合作，外出到墨西哥灣裡去打大魚！」老漁夫先生快樂極了。

不知道老漁翁與大帥哥究竟出去打了幾次漁，老漁翁居然在附近的七河醫院急救不及而去世了。

如玉與老漁翁先生已經結婚 54 年，早已習慣彼此相依為命，如今突然失去老伴，如玉覺得生活非常之不完整，幸好有鄰居大帥哥艾迪經常前來照顧，她才得以勉強存活下來。

「我想把老漁翁的漁船賣出去，免得睹物思人，只要見了這艘船，我簡直活不下去了！」想起老伴兒的萬種體貼，千般溫柔，如玉涕淚漣漣的說。

「那我叫第 19 號公路旁邊的那一家代理售船公司把你家的船拖到他們的售船廣場，公開展售，那裡能見度比較高，主顧比較多，且又免得你家有閒雜人來客去，對你比較不方便，也不安全！」哇，新鄰居艾迪哥真是太貼心了！

「既然沒有船了，不如把空船蓬一併賣掉吧？！」如玉問。

「船蓬跟船隻不一樣，不能拖來拖去，只好找人來現看現買。」帥哥艾迪想了一下，如此回答。

數天之後，果然有一位叫做哈里森的南方紅脖子老粗，開了卡車，帶了工具，把船蓬拆下來帶走了。

「哈里森先生，你要拆干净一点，因为我没有工具啊！」如玉說。

「哦，．．．妳沒有先生嗎？妳一個人嗎？」哈里森很感興趣的問。

那知隔壁的帥哥艾迪由他自己的陽台上，目送哈里森卡車出去之後，立刻跑了過來找如玉說話。

「如玉，剛才我看見哈里森找你講話，講了很久，說些什麼？」艾迪帥哥很不放心的問如玉。

「哎，他隨便問了一些我的情況！」如玉不經意的回答。

「你可要小心喲，哈里森是我們附近名譽很壞的混混，你真的要小心！」艾迪　很嚴重地警告如玉。

果然，第二天早晨，哈里森踩著他的大皮靴，噔噔的又上了如玉家的二樓。

「喂，如玉，你泡杯咖啡給我喝罷!」哈里森坐在如玉家的客廳涎著臉請求。

如玉嚇了一跳 不知道是不是應該泡杯咖啡給他, 正在此時，如玉家的門鈴響了。如玉慌忙下去開門，一看按門鈴的是艾迪，馬上放下了一顆忐忑的心，煮了兩杯咖啡給客人們飲用。

這樣做了數次之後，混混哈里森只得知趣地放棄了騷擾。

不久，如玉開車出了車禍之後，暫時沒有車開，小艾自動上門問如玉需要不需要搭他的便車到超市中去買菜，為了感謝他的好意，每次買完菜，如玉就請小艾出去吃午餐。

「哎，如玉，那邊有個金髮美女一直看你，你認識她嗎？」正在吃午餐的艾迪突然停上咀嚼，問坐在對面的如玉。

「艾迪，怎麼可能有金髮美女看我？人家是看你帥哥的吧？」如玉一面回答一面轉過頭來一看，果然是她認識的人。

「哈羅，麗莎，你今天沒有去上班嗎？」原來是小鎮上郵局工作的工作人員麗莎，因為不久前如玉想要重新申請一個中華民國護照，護照上需一張新的照片，是麗莎替如玉拍的照片，所以麗莎可以說是如玉認識的人。

「我下個月就要由郵局退休了，退休後要找一個便宜一點的房子，距離郵局遠一點沒有關係。」麗莎回答。

「如玉，妳的樓下的房子不是空的嗎, 租給麗莎住怎麼樣？」帥哥艾迪很熱心地問。

「租給麗莎住？當然好，可是我打算搬到靠兒子近的加州去住，正在賣房子，房子賣了怎麼辦？」如玉有點猶豫。

「現在還沒有賣掉，等到賣掉了再說吧！」帥哥艾迪說。

後來，麗莎果然搬到如玉家的樓下去住，兩個女人正好作伴。

如玉賣掉房子之後，果然如願搬到加州去啦。她聽帥哥告訴她說：新的房主不願意把樓下出租，那丽莎就只好搬到艾迪的房子的樓下去住了，後來，艾迪的父母搬到艾迪的樓下暫時居住，麗莎就自己搬到樓上去與艾迪同住了。

艾迪就此做了丽莎的終生護花使者。

Remove the Clouds and Let the Sun Shine Through

--Edited by Sandra Chin02/25/2017

Old man Zhang wears a pair of glasses that have large circles encircling smaller circles because his elderly eyes suffer not only from deep myopia but also with severe cataracts.

Early one morning when he was outside， a Chinese beauty in bright red and green greeted him from far away. "Good morning， Teacher Zhang! Are you going out for a walk?"

"Good morning! Good morning to you too!" The old gentleman stopped in a hurry and bowed to the shadow in front of him repeatedly. We can never blame anybody who is too polite， can we?

"Teacher Zhang， in front of you is a tree; I am behind you， " the woman said.

"Who are you? And where are you?" The old gentleman made a quick turn and nodded randomly toward the direction of the sound.

"Teacher Zhang! I am your neighbor from across the hall， Big Sister Chu!" This is one of the good Chinese customs： call any woman who is older than yourself "big sister" instead of "old lady" so you will not get into trouble.

"Ha， ha， ha!" the two old neighbors laughed.

During the Spring Festival dinner， a giant plate of cold cuts was served. In order to show respect to the elderly， we all said， "Mr. Zhang， please， you take the food first!"

"Well， good， good， " said old gentleman Zhang. He tried to pick up the red shrimps with his chopsticks， but to no avail.

"Hey， Teacher Zhang， the shrimps are in the middle of the plate. What you tried to pick up are the red prawns painted on the side of the plate!" Big Sister Chu whispered to him.

*****

One day， Xiao Li and Xiao Wang went to visit their old neighbors. Thou shall be nice to your neighbors， right?

Entering the house, they saw the old man was wearing his glasses, but his nose was sticking to the newspaper and there were several large and small magnifying glasses placed near him on the side table.

"What's happened? Does Teacher Zhang smell some weird odor from the newspaper?" Xiao Li asked with surprise.

"No! Recently the elderly gentleman's eyes are getting worse. Even when he sticks his face right up to the newspaper, he still can't see the characters with more strokes, " explained Big Sister Chu who was sitting beside him.

Pediatrician Xiao Wang suggested the old man see his colleague, Dr. Tsai, for a physical check-up. The examination revealed the old man's cataracts were worsening so he reluctantly agreed to surgery.

" Many years ago, ophthalmologists were afraid my myopia was too deep and there was a risk of blindness resulting from retinal detachment, " old man Zhang worried.

"Now , with the development of modern medical techniques, there is more advancement in surgical technology than a few years ago. So long as the ophthalmologist says there is no problem, you have nothing to worry about, " Xiao Wang comforted old man Zhang.

The right eye was more serious so surgery was scheduled for July 2. Two weeks before the surgery, old man Zhang had a full body check-up. His height, weight, blood pressure, and EKG were all normal. He didn't even suffer from osteoporosis! On the scheduled day, we all waited in the waiting room. Dr. Wang accompanied the elderly gentleman into the operating room where the nurse took over. After about 45 minutes, we saw the old man walk out with a smile on his face.

"Why did you come out?" Big Sister Chu asked with concern.

"Let's go home!" old man Zhang said.

"What?"

"The nurse put some drops in my eye, like an anesthetic, and then told me to look at the light for half an hour. Then Dr. Tsai said the cataracts had been removed, " old man Zhang recalled. "I can go home now!"

"Dr. Tsai inserted a new lens into Mr. Zhang's eye, and now he not only does not have cataracts, he is no longer near-sighted either!" Dr. Xiao Wang explained.

"He doesn't even need eye patches? Is there anything different?" Big Sister Chu was a little bit worried.

"Now, my right eye sees a light aperture, " Mr. Zhang announced.

Fortunately, after one night's sleep, the aperture disappeared. When old Zhang walked out of his bedroom, he was startled to see a good looking elderly lady in his living room.

"Who are you? What are you doing at my house? How did you get in?" the old gentleman asked repeatedly in surprise.

"Teacher Zhang, I am your old neighbor from across the hall, Big Sister Chu! YOU are the one who invited me over, and I have the key to your door!" The stylish old lady was also shocked.

"Well, then, who is the handsome young man with freckles? And why is he playing with my grandson?" The old man still did not feel at ease.

"The handsome young man with freckles is my grandson David Chu. He is your grandson Bill Zhang 's classmate; they play with each other every day!" Lady Chu replied.

Didn't the doctors claim that the operation would improve his vision?

"Well, before the surgery, my eyes were youthful eyes; the world was a splendid place. It was like a fog, like rain, also like flowers. Now my eyes are magic mirrors for revealing goblins, rouge painted on ladies' cheeks, wrinkles on old friends' foreheads. The acne on the faces of the youths and the freckles on children's noses are all perceived very clearly through my new eyes." The old man nodded his head and announced with a smile.

That was what had happened!

"Oh, now you can see the wrinkles on my face? Is that good or bad?" Lady Chu asked anxiously.

"This is good, of course, it is so good. This is what we

call old friends!" The old gentleman was very satisfied.

The medical expense of surgery for one eye was $8, 131.39 US dollars, but only $610.22 was out of the individual's own pocket. The rest was paid for by Medicare. Therefore, Mr. Zhang had the cataract removed from his left eye one month later.

All the elderly people from the senior citizen center heard that the confused old man Mr. Zhang's vision had become so good, and that Medicare paid 92% of the surgical expenses. So they all decided to have their eyes checked and their cataracts removed as well. It is a waste not to use senior benefits.

"Since the result of your surgery was so good, I want to have my cataracts removed too!" Big Sister Chu told her old friend Mr. Zhang.

"Old fish with pink gills swims from muddy water to the clear stream; it moves its head, then moves its tail, it moves its head..." the old gentleman could not help but sing out very happily.

One day, old man Zhang knocked on his neighbor's door. Seeing that the one who answered the door was Big Sister Chu herself, the old gentleman quickly stepped forward.

"Big Sister Chu, It took me a lot of time and effort to buy two tickets to 'Dream of the Red Chamber' at the San Francisco Theater. Let's go together to watch the show tomorrow, " old man Zhang invited Big Sister Chu in all sincerity.

"Does it have to be tomorrow?" Big Sister Chu answered hesitantly.

"Yes, I was in a long line for a long time to buy these tickets, and the only ones available were for tomorrow!" answered the old gentleman.

"I am so sorry, but tomorrow is the day I have scheduled to remove my cataracts!" Big Sister Chu told her old friend Zhang.

撥開烏雲見太陽

張老先生戴着一付大圈圈裡面有着小圈圈的眼鏡，因為他老人家雙眼不但有有深度的近視，也有很嚴重的白內障。

清晨出門，一位穿紅著綠的中華儷人遠遠地向他打招呼：「張老師早！張老師出門散步嗎？」

「妳早，妳早！」正在行走的老人家慌忙停步，頻頻向前面豎著的影子點頭打招呼，禮多人不怪嘛。

「張老師，您的前面是一棵樹。」那位東方女人指出來。

「妳是誰？妳在那裡？」老先生連忙轉了一個身，向聲音發出來的地方胡亂猛然點頭。

「張老師!我是您家對門的朱大姐呀!我在您老人家的後面呢!」還是咱们中國習慣好，凡是比自己大的女士，一概叫大姐，　绝不惹上麻煩。

「哈，哈，哈。」两位老鄰居都笑了起來。

春節聚餐時最先上了一個巨型的冷盤，為着敬老尊賢，大家都說：「張老先生先請！先請！」

「好，好，好。」老人家當仁不讓，舉起筷子，對著盤子四週的大紅蝦挾了好幾筷，可惜什麼都沒有挾起來。

「喂，張老師，菜在盤子的中央，你挾的是盤邊畫的紅色大蝦呀！」坐在老人家身边的朱大姐只得俯在他耳边提醒他老人家。

一天，小李和小王去拜訪年老的鄰居，和親睦鄰嘛！進門一看，只見張老師戴著近視眼鏡，鼻子貼著報紙，一旁的書桌上大大小小擺着好幾付放大鏡。

「怎麼，張老師聞見報紙有什麼怪味嗎？」小李很詫異地問。

「不是啊!近來老人家眼睛愈來愈不行，就算臉蛋貼著報紙，筆劃多的字仍然看不清。」坐在一旁的朱大姐向大家解釋。

小兒科医師小王建議帶老先生去他的同僚蔡眼科那裡去檢查。檢查的結果，說張老師近來白內障日益加深，老人家勉強同意開刀。

「不知疼不疼，效果好不好？多年前眼科医師怕我近視太深，開白內障有視網膜脫離而導致失明的危險。」張老先生十分担心。

「現在医學日新月異，目前外科技術比數年前進步太多了，只要眼科醫師說沒有問題，老先生就不必担心啦。」醫師小王安慰老人家。

右眼比較嚴重，擇定七月二日開刀，二週前要做全身檢查，身高、體重、血壓以及 EKG 一切正常，連骨質都沒有疏鬆，開刀的那天，大家動員到蔡眼料的候診室等候，小王醫師陪著老人家進入手術室，將老先生交給護士，大約三刻鐘左右，只見老先生笑嘻嘻地由裡面出來。

「出來做什麼？」朱大姐迎上去，関心地問。

「回家罷。」張老先生說。

「什麼？」

「只在右眼裡滴了像眼药一樣的麻药，然後叫我眼睛對著光半小時，蔡医師說白內障已經除去，可以回家了！」張老先生回憶說。

「蔡医師替張老先生換了一個新的晶片，他老人家不但沒有白內障，也已經不再近視了。」小王醫師解釋道。

「連紗布都不用包嗎？有沒有什麼異樣呢？」朱大姐有點不放心。

「現在，右眼看見一個光圈。」張老先生向大家報告。

好在睡了一覺，第二天起床，光圈就消失了。老先生走出臥室，看見客廳中坐着一位不認識的漂亮老太太，著實嚇了一大跳。

「請問妳是誰?在我家幹什麼?怎麼進門的?」老人家連聲問道。

「張老師，我是您老人家對門的鄰居朱大姐啊!不是你邀請我過來的嗎?我有您家開門的鑰匙呀!」那位時髦老太太也嚇了一大跳。

「那麼，那位有雀斑的小帥哥又是誰?他為什麼跟我的孫子在玩耍呢?」老人家仍然不放心。

「那位有雀斑的小帥哥是我的孫子朱小明，他是您孫子張小傑的同學，他們不是天天在一齊玩的玩伴嗎?」朱大姐回答。

奇了怪了，不是說動手續去掉白內障對於視力有幫助的嗎?

「唔，開刀以前我的眼是青春眼，花花世界一片迷迷糊糊，像霧像雨又像花，現在变成了照妖眼，女士們塗的胭脂花粉，老朋友額上的皺紋，青少年臉上的青春痘，小朋友鼻边的雀斑，在我老人家的法眼照視之下，一概清清楚楚。」老人家點點頭，笑著宣佈道。

原來如此!

「哦，您現在看見我臉上的皺紋了?那是好還是不好呢?」朱老太太慌忙問道。

「這樣好，當然是這樣好，這才是名符其實的老朋友嘛!」老
先生非常滿意。

開刀無病無疼，開一隻眼的医药費用共八仟二佰卅一點卅九
美元，自己只付六百一十九點二二美元，其他的全由社会保健支
付，第二个月，張老先生就將左眼的白內障也除去了。

老人中心的老先生老太太們聽說以迷糊著稱的張老先生的
視力变得如此之好，加以社会健康保險付了百分之九十二，全部
決定都要去檢查一下双眼，若有白內障一定要去開刀，老人福利
不用白不用。

「老老魚兒粉紅鰓，濁流游到清溪來，頭動尾巴擺，頭動尾
巴擺．．．。」眼聰目明的老先生忍不住開口唱道，狀至快樂。

「開刀既然這麼好，我也想要去把白內障去除了!」朱大姐對
老先生宣佈。

一天，張老先生到對門朱家敲門，老先生見應門的是朱大姐，
連忙一個箭步搶上去。

「朱大姐，我特地花了九牛二虎之力到舊金山大戲院去搶購
了二張”紅樓夢”的戲票，咱們明天一同去觀賞如何?」老先生
喜孜孜地邀請朱大姐。

「一定得明天嗎?」朱大姐十分遲疑。

「是啊，排長隊搶購到的是明天的票!」老先生回答。

「可是，我也要開除白內障，開刀的日期是訂在明天啊!」朱
大姐告訴她的老朋友張老先生。

A Tomb-Sweeping Festival Story

Edited by Descartes Li

When I was young, every Tomb-Sweeping Festival, my parents always burnt three sticks of incense, prepared a small table of chicken, pork, fish or other rarely served good dishes, filled up three bowls of white rice, and placed a pair of chopsticks at the side of each rice bowl.

One day, I asked: "Why three bowls?" My father bowed three times first then explained to us: "One bowl is for my father, your late grandfather, of the Yu family. Another bowl is for your late grandmother, who was my official mother, Lady Xu. The third bowl is for my real mother Huang." Let me tell you the story of my real mother.

My biological grandfather, the father of my biological and real mother died early. This left my mother to bear the responsibility of supporting her mother and her younger brother. She became a domestic helper to our Yu family at the early age of 13 and became known in the Yu household as Little Huang.

One day, Lady Xu, your grandfather's first wife, took pity on the hard life of Little Huang. Years earlier, Lady Xu herself had come to the Yu family, with only a dowry chambermaid to accompany her. She sent a small bag of coppers to the kitchen to give to Little Huang. Little Huang took the small bag of money, hurried back to Lady Xu's rooms and said to the mistress: "Madam, you have already paid me! I could not accept this extra money!"

"Since you don't want the monetary reward, is there something else you want?" Lady Xu, who just recovered from an illness, asked with a kind smile. The Clever little Huang replied: "Well, my brother hangs around with the local rascals all day. It really is not good for him. Is there any way that he can go to the the home of the Yu family tutor? He can help with chores, and maybe learn to read and write some simple characters?" Our docile, kind-hearted, Xu grandmother laughed."You are not so stupid, to exchange such a small reward for a job for your younger brother!"

After the death of childless Lady Xu, the Yu family was in decline. The elders all felt that the successor wife should be a lady came from the family with an equivalent (or higher) social status. Little Huang was just the servant in my father's bedroom. This argument dragged until I was born. Since I was the male heir, your grandfather made Little Huang girl his official wife. Little Huang was no longer a little girl, she was now Lady Huang.

At the same time I was born, my uncle, the little rascal Lady Huang's younger brother hung around with was arrested by the police. But Lady Huang's younger brother knew how to read and write, and once word was sent that he was related to the good Yu family, he was sent to the police academy. But my father was sick, and as part of the deal, the Yu family decided to break up.

My uncles, taking advantage of your grandfather's poor health, and his new wife's young age, divided up family properties, and your grandparents only got a a house and some land which were already pawned out to the local Zhou family.

My mother Huang was young, but smart and vigorous. On our lands, she worked with the tenant workers during the day, and even slept in the field grass hut with the women workers at night. My father died one year later, but my mother succeeded in paying off the house to the local Zhou family.

Right after the death of my father, the Zhou's claimed the house where we were living, and moved in. Every mealtime, the Zhou's, young and old, rushed to the dining table, and their chopsticks always reached the food first. My mother took the matter to the higher officials, but the court refused even to hear the young widow with a fatherless toddler in her arms. That toddler was me.

What did the helpless young widow Huang do? She figured out a way. One chilly dark night, she slipped into the backyard, from under the eaves pulled a broken brick, and used it to knock her own forehead.

"Help, Help!" The young widowed mother cried at the top of her lungs, and her shouts were heard by all the neighbors. At the county court trial, the judge/mayor saw in front of him a little boy kneeling along with his widowed young mother, her beautiful face was covered with blood.

The judge exclaimed: "What happened?" The Widow Huang immediately took the written contract out for the authority to see. And the judge ordered on the spot that the Zhou family should move out Yu's house immediately, never again harassing the toddler orphan and the young and beautiful widow.

Afterwards however, your grandmother still feared the persecution of the Zhou's, and so she took me and we left our hometown village Xinghua, moved to Yangzhou, a bigger city. We moved near the residence of one of my father's aunts, whom we called Second Grandmother Yu.

"You will spoil your son by sending him to school!" Said some relatives. At that time, the typical custom was to use opium smoking to keep first sons at home, but your grandmother had a different idea. She sent her only son, the original bowing, orphan toddler, to far away Shanghai to go to university.

I think I lived up to my mother's expectations. After four years of university, I graduated summa cum laude and won a 400 silver dollar prize. However, after I came home, I brought an even more surprising news to my mother. I met a young woman, Miss Chen, who came from Yancheng, which means Salt City. She had come to Shanghai to receive University education in Shanghai. Your grandmother had meant to use a matchmaker to find a good daughter-in-law for herself, but now, her son, had found himself a new modern woman with higher education. I married Miss Chen, your mother, in Shanghai. At first, my mother did not agree, but soon changed her mind.

Your grandmother realized that instead of making her son sad, it is better to add a beloved daughter-in-law. So she treated my love like her own daughter, taught her to light the fire for cooking, do the laundry as well as the sweeping, and so on. She also instructed my wife, formerly Miss Chen, on how to treat servants, so that she would gain their respect and not be manipulated. At the time, I believed my considerate and smart young wife was learning these very un-modern skills out of respect for my mother.

During this time your grandmother used to say in front of people that she could not find her glasses, do let Miss Chen read the newspaper for me. Just to show that she had a daughter-in-law who could read. This not only showed off that our mother was a filial and educated woman， but also masked the fact that grandmother herself was illiterate.

I thought my life was settled. I was going to be the local high school mathematics teacher and live with my lovely intelligent wife. But I had not expected that the Sino-Japanese War would break out! We fled to Chongqing， with all the housework having to be done by the two of us. If not for your grandmother's teaching， we could not have survived those days of World War II! Your mother always praised her mother-in-law for her magnanimity and cleverness.

I was only a little girl when my father told me this story. He is now long gone. One day I too will be gone and maybe on some Tomb-Sweeping Day in the future， you can tell you granddaughter about your grandfather and his real mother.

重振家業的祖母

　　小的時候，每逢清明節，我的父母都要準備一個小桌上面供了一些魚肉雞等平常吃不到的好菜，然後添三碗白飯，碗邊安放三雙筷子。

　　怎麼三碗啊？我們問。

　　「一碗是你們祖父的，一碗是大祖母許氏的，第三碗是給我的親生母親黃氏。」爸爸鞠完三個躬之後解釋道。

　　據父親告訴我們，親生黃氏祖母的父親很早就去世了，所以黃家小丫頭才十三歲，就在我們余家幫傭，擔當起養活自己母親和弟弟的責任。

　　一天，祖父的元配大家小姐徐氏派了她陪嫁的小丫頭拿了一小袋銅板到廚房，賞給臨時僱來幫忙的鄰居黃小翠。

　　黃小翠拿了賞錢，匆忙趕到徐氏的病房，向主母推辭道：「太太，你已經給我工資了，服侍您吃飯給您打掃房間，是我份內的事喔！怎広還能另收賞錢呢？」

　　「你既然不要賞錢，那告訴我，你想要什麼呢！」大病初愈的許氏笑嘻嘻地問。

　　「這樣吧，我的弟弟整天跟巷口的小流氓混，實在不是一個辦法，不知道可不可以到余氏家族私塾老師家去打雜，他如果因此能認一點字就好了！」聰明伶俐的小丫頭黃小翠回答。

　　「你倒不錯，用這麼一些小賞錢，竟然想替你弟弟弄一份差事！」心地善良、性格溫順的徐氏祖母笑了起來。

　　沒有生育的大祖母許氏過世以後，祖父家中饋乏人，而余氏家族雖已日趨式微，可族人卻仍然認為美麗健康、聰明伶俐的小丫頭只能做祖父的房中使喚，他的繼任夫人應該是一位門當戶對的小姐。這件事情一直拖到我的父親出世，祖父才以有了兒子繼承人為由，毅然決然地把祖母正式收為填房，那時黃家巷口的小混混已被警察局抓去，小舅公因為能識之無，就被送到警察學校去學習，訓練成一個為民除害的警察學員。

　　余家族人因為欺負祖父身體不好，填房祖母年輕，加以黃家沒有有力的父兄撐腰，所以分家的時候，祖父母分到的家產全是已經典當出去的田地房產。

　　祖母不願意讓余家族人有更多藉口，給祖父添加更多的麻煩，就毅然決然地承擔當起贖回已經典當出去的房地產的重任，每天辛苦持家，任勞任怨，白天與佃戶長工及女傭們一同工作，甚至晚上還與與女工們睡在田邊草寮中。

　　因為祖上缺錢，他们家的房子曾典押給周姓人家，後來祖父母湊足了銀兩，終於將房屋贖回自住，那知祖父一過世，周家突然舉家老少，全部搬到孤兒寡婦家中，吃飯時先上桌坐，上菜時先伸筷挾，而且揚言要賴著永遠不肯走了，余家狀子告上衙門，竟如石沈大海，遲遲不見處理，年青的黃氏無奈，眉頭一皺，計上心來，乘夜晚月黑風高之際，溜到後院，由屋簷下抽出一塊破磚朝自己的額頭上敲去。

　　「救命呀，救命呀!」年輕的祖母高声吵嚷起來。

　　她的喊聲一時驚動了鄉親地保街坊鄰人，縣衙門立刻升堂，只見帶了幼子跪在堂下寡婦楚楚動人的臉上蓋滿了鮮血，主審敲響驚堂木，黃氏立刻由懷中取出住房銀貨兩訖的合約，看得主審的縣長怒由心起，也不及問起這位少婦為何受傷，當場下令要周家全部立刻搬出，永遠不許再騷擾余家尚未成年的孤兒以及年青美麗的寡婦。

　　事後，祖母怕族人欺負年幼的父親，受了時任江蘇省民政廳長，後來在台灣做過行政院長的叔公余井塘母親的鼓勵，毅然絕然地帶領了女傭及長工們離開家鄉興化，搬到揚州，在叔公余井塘母親家的附近居住。

　　「慣兒不慣學!」祖母說。當時的社會風氣是要用吸食鴉片煙來繫住兒子們，可是那時我的祖母卻獨排眾議，把她的獨子送到上海去讀暨南大學接受新式教育。父親也不辜負祖母的希望，四年之後，以全校第一名得到了四百枚大洋銀元的獎金畢業。

　　不過呢，我父親回家以後，卻帶了一項驚人的消息，告訴祖母他在上海認識了一位由鹽城到上海讀書的程家千金小姐，一直央求祖母找人到母親家去提親，本來祖母的意思是要用父母之命媒妁之言給自己找一個好媳婦的，現在，父親居然自己找了一位受過高等教育的新式女子，起初老人家雖然非常不樂意，但是很快就改变了主意。

　　「按說，我不是她親自選擇的媳婦，老人家認為與其使兒子傷心，不如增加一個自己疼愛的媳婦，不但沒有虐待排斥我，反而待我有如親生，由生火煮飯、洗衣掃地等教起，乃至如何管理傭僕，婆婆的原意是教導媳婦應該懂得如何親自操作家務，才不會被手下傭人們拿捏，那知中日抗戰發生，我和你父親逃難到了重慶，所有家務都得親自動手，如果沒有你們祖母的教導，以我一個在程家四體不動，五穀不分，到上海去讀書的千金小姐，哪裡能夠挨過抗戰時代那些貧困的日子!」母親也常常在父親及我們面前稱讚祖母的大度和聰明能幹。

　　因為我的父母親經常交相稱讚的緣故，所以雖然這輩子我從來沒有見過她老人家，但是我第一個最愛的女人就是我爸爸的親生媽媽，也就是我一心嚮往的祖母。

「你们的祖母常常當着人面說不必找眼鏡了，叫秀山(母親的小名)替她讀報紙罷!」父親說，這樣不但張顯了我母親是一位孝順且受過新式教育的女子，而且還掩蓋了祖母自己本身不識字的事實。

不認字?那她怎麼管理地主家的家庭財務呢?我們又問。

「你们的祖母是一位絕頂聰明的女子，我們中國不同面額的鈔票是不同顏色及大小的，她就根据紙幣的顏色和大小自己發明了一套查帳的方式來監督帳房的帳薄。」媽媽告訴我們。

那，如果她到了只有同樣大小綠色鈔票的美國，不就沒了徹嗎?我們再問。

絕對難不倒她，聰明的老人家一定另有一套格外高明的辦法來解決這個問題的!對於這一點，我爸爸媽媽異口同聲一致非常肯定。

Here, There, Everywhere Are All Suitable for Doing Business

Edited by Pearl Karliner-Li(3/24/20)

The first time I met Wong Ahjei (elder sister) was at a flea market in central Florida. She was doing business there.

"It's $39.99 per bag， all genuine leather hand bags!" A woman from a stall shouted in English with a warm Taiwanese accent; "If you pay cash， you don't have to pay the tax， I'll pay it for you."

Hearing that accent， I turned around， and saw a large group of white American women swarmed around a middle-aged Asian woman， each of the white women snapping one or more purses.

"Are these genuine?" A white woman holding a purse in her hand asked very seriously.

Hurriedly， I tried to squeeze toward the perimeter of the group， hoping to find out how she would answer.

"Of course， they are all the genuine imitation "'B-grade' goods..." The Chinese woman smiled and replied in English， "Do you want the better genuine imitation 'A-grade' goods? I do have them too， in the storage house next to me." There， next to her stall， stood a small storagehouse.

"Can I go in and have a look?" The potential customer asked.

"As long as you are sincerely planning to buy， of course you can!" The Chinese woman from Taiwan smiled and said， immediately pulling out the key to the storage house from her purse.

When the door of the warehouse was opened， four middle-aged white American men in civilian clothes suddenly appeared. Three of them pushed into the warehouse， taking all of the real "A-bags， " while the fourth one swept all of the "B-bags" from her stall.

"Please come with us to our office!" The man leading the group said; it looked like she had to go with them.

"No English， I don't know English!" She said in stuttered

English.

"Do you want me to be your translator?" Standing next to her, I saw she needed help very badly, so I instinctively volunteered to do the translation for her.

"No English, I don't speak English!" she continued, but waved her hands at me to indicate that she didn't want me to interfere.

I thought about it, and suddenly realized that her English could not be poor. Wasn't she communicating and doing business with a lot of white American women in fluent English only a minute ago?

The four men discussed privately. "Since this woman can't speak English, and thus can't be punished, taking her to the office doesn't serve anything, it's better for us to just confiscate these counterfeit goods." The men took all the "A" and "B" goods and, like a gust of wind, they went away.

I waited for these four men to disappear completely, then smiled and said to her, "Hey, you're so smart!"

"Am I? Am I really smart?" She wondered with a smile.

"Otherwise you'll have to go to their office with them, and do you know what their office looks like?" I was afraid that she might be tortured.

"It's good not to go, and we don't want to know what their office looks like." she concluded. "My name is Wong AhJei (elder sister) from Taiwan, what's your name?" She asked me.

"My name is Gwen Li, also from Taiwan." And with the exchange of our names and origins , us two fellow countrywomen felt much closer.

"Hey, Gwen Mei (younger sister), " Wong AhJei suddenly thought about something. "I'll take you to an office that I guarantee you will like to see."

"Okay, when?"

"Now, of course it is now!"

Sure enough, she asked me to help her pack up her stall, instructed me to sit next to her in her pickup truck, and started the engine. On the way toward the city of Tampa, we chatted very happily and forgot the time. Before I knew it, her pickup stopped in front of an unremarkable house.

By the time we stepped out of her truck, we were already familiar and as intimate as old friends.

In the doorway we were received by a Latino man and woman, who seemed to be a couple. Wong Ahjei told them what had happened that morning in the flea market. She told them about the four men and how they had confiscated all of her "A" and "B" goods.

"I'm keeping my mouth shut, I haven't said anything! So I hope you will return all the money I paid you!" Wong AhJei spoke politely and took out a receipt.

The couple's facial expressions turned immediately from polite to grave. The woman took the receipts from Wong Ahjei's hands, stood up and they both went to the back room. After a while, the two returned from inside and gave Ahjei a thick envelope stuffed with paper money. Then, very enthusiastically, they brought out piles of Mexican food from within and the two of us enjoyed a very delicious hearty Mexican lunch.

"Ahjei, your leather bags were confiscated by the people at the flea market, why are these two willing to reimburse the money to you, ah?" I could not help but ask as we drove away.

"Hey hey, they're the wholesaler who sold the merchandise to me, I didn't tell those four people about them, I didn't reveal where their base is, now I just want my own money back, and didn't blackmail them - I've done my best for them!" She very much appreciated her own actions.

"Hey, Wong Ahjei, you're amazing!" I sighed and laughed.

"Am I?" As she drove, she laughed happily too.

The two of us laughed loudly together. As we were feeling smug and satisfied, we were overwhelmed by a roaring and flashing and noticed we were being followed by a police car. So, we slow down and stopped our pick up at the side of the highway.

One of the two policemen knocked on her window, and Wong AhJei had to open the window, and the other officer wrote down her driver's license and license plate number on a piece of paper and handed it to her.

"What's the matter?" I hadn't figured out the situation yet.

"We were just so smug，only paying attention to our own talk，and I drove too fast. I got a ticket for speeding." Wong Ahjei's face turned pale.

"Well，how about I pay half of the amount on the ticket，because there were two people talking to each other. It's not only your fault，" I comforted her.

"No，I can't!" She said with a sadder expression.

"Why not?" I didn't understand.

"Because I've got too many tickets! This last one has increased my violation score significantly… As punishment，my driver's license will be revoked，that is to say I won't be allowed to drive for a whole month."

"Is there any way to resolve it?" I asked.

"I can go to traffic class. This would reduce some of the fines and some offending scores." It turned out that Wong Ahjei was very clear about the traffic rules.

"Well，this next a month，I'll drive you to classes!" I said very bravely.

On the first day of the traffic class，I took her to class. She left my car carrying a huge bag full of things.

When the time came for me to pick her up，I sat in my car and waited outside for a long time. When I didn't see her coming out I parked the car and walked toward the classroom.

From the corridor I saw a lot of people gathered together. As I reached the periphery of the crowd，I found that she was busy doing business with the other men and women taking the traffic violation course. Everyone was scrambling to buy pen and paper，stationery，and books from Wong Ahjei，there were also some women trying to purchasing the lipstick，rouge and perfume from her. The last blond woman even bought an emerald-green jade bracelet from her.

"The beautiful blond who bought the jade bracelet is our class teacher，so I had to give her a price break." She explained to me.

Ah，for Wong Ahjei，here，there and everywhere are all suitable for doing business!

此處，彼處，處處可以做生意

　　我第一次看見汪阿姐，是在跳蚤市場上，她正在做生意。

　　「全部$39.99一隻，全部是正貨喔！」一個攤位上傳來一位女子用英語吆喝的聲音，而這聲音帶著親切的台灣口音。「用現金買的話，　我付稅金，　您不必付稅哦!」

　　定睛望去，只見一大堆老美鄉下白女人圍著一位東方面孔的中年婦女，人手一隻或多隻皮包，正在搶購。

　　「你這是正貨嗎？」有老美鄉下女人，手中舉著皮包，很認真地問。

　　我連忙也努力擠進外圍，且看她如何回答。

　　「當然是真正的仿製正B貨！」這女人笑嘻嘻地用英語回答，「你要比較更正的仿製 A 貨嗎？我也有，就在旁邊的庫房內。」攤位旁邊有一個小小的倉庫。

　　「可以帶我進去看看嗎？」白女人問。

　　「只要你有誠心要買，當然可以！」這位台灣來的中國女子笑著說道，並且由掛在身上的皮包中掏出鎖匙，去打開倉庫的門。

　　當倉庫的門被打開的時候，突然出現了四位穿了便服的美國中年白人男子，三人搶進倉庫，拿了很多真正的 A 貨皮包，另外一人把她攤位上真正的 B 貨皮包一起收刮了。

　　「請你跟我們去我們的辦公室一趟！」這四人中為首的那人說道，由情勢看來她是非去不可了。

　　「No English，I don't understand English !」她結結巴巴地說。

　　「妳要我做翻譯嗎？」站在旁邊的我看她這樣，連忙用中文自告奮勇要替她做翻譯。

　　「No English, I don't speak English ！」她的口中繼續說，不過卻私地下對我搖手不希望我參予。

　　我轉念一想，突然恍然大悟，她的英文怎麼可能差呢，剛才不是用很流利的英文跟一些老美鄉下女人交流做生意的嗎？

　　「這女人既然不會英文，又不能用嚴刑拷打，我們帶她去辦公室也沒有意思，不如把手上的這些冒牌貨貨拿去交差就算了。」這四個人私下商量之後就拿了全部真正冒牌的 A 貨及 B 貨，一陣風走了。

　　我等這四人走遠了，笑嘻嘻地對她說道：「嘿，妳真聰明！」

「是嗎？聰明嗎？」她也笑嘻嘻地承認了。

「不然你就得跟他們去他們的辦公室了，不知他們的辦公室什麼樣子？」我說。

「管他什麼樣子，還是不去的好！」她下了一個結論。

「我叫汪阿姐，台灣來的，妳叫什麼名字？」她問我。

「我叫余國英，也是台灣來的.」老鄉互相交換姓名之後，覺得親切起來。

「喂，國英妹，我帶你去一個辦公室，保險你喜歡去。」她突然想起來。

「好呀，什麼時候？」

「現在，當然現在！」

果然，她要我幫她把他的攤位收拾收拾，讓我坐在她旁邊，就開了她的小貨車，向天柏市方向走。路上我們談天說地得非常起勁，所以很快就到了一座不起眼的房子前面。

停車的時候，我們已經熟悉及親熱得好像多年的老朋友一樣了。

進門之後，有一對像夫妻一樣的兩個拉丁裔出來接待我們，汪阿姐把今天上午在跳蚤市場有四個人來將她真正的 A、B 貨全部充公的事情，前前後後用英語全部告訴了這兩個墨西哥人。

「我這是守口如瓶，什麼都沒有說喔！所以我希望你們把我付給你們買貨的錢全部退還給我！」汪阿姐拿出一張收據，很客氣的說。

這兩位墨西哥人立刻臉色凝重，由阿姐手中取過收據，進到房內，過了好一陣子，兩人才由裡面出來，不但手中拿了一個厚厚實實裡面裝滿了錢的信封，並且很客氣地由裡面一一端出很多好菜，請我們兩個人吃了一餐非常豐盛的墨西哥午餐。

「阿姐，你賣的皮包被人家拿去充公了，這兩個墨西哥人為什麼肯把錢賠給妳啊？還請我們吃墨西哥大餐！」回程的路上，我大驚小怪地問。

「嘿嘿，他們兩人就是墨西哥批發商，我沒有把他們的事情告發，也沒有洩漏他們的大本營在哪裡，現在只向他們要回本錢，沒有敲詐他們，對他們已經算仁至義盡了！」她對自己的行為十分欣賞。

「哎，汪阿姐，你太了不起了！」我嘆了一口氣，當面恭維她之後，笑了起來。

「是嗎？」她一面開車，一面也高興地仰天哈哈大笑。

我们两人一同縱聲大笑，正在得意忘形的時候，一輛警車呼嘯而至。

兩位警察中的一人敲她的窗子，她只得將車窗打開，另一名警察把她的駕駛執照，車牌號碼，寫在一張紙上，把這張紙交給她之後，兩人揚長而去。

「怎麼一回事？」我還沒有搞清楚情況。

「剛才太得意，太專心講話，車速開得太快，吃了一張罰單。」阿姐突然臉色大變。

「那這樣好了，罰款我也幫你出一半，因為是兩個人在講話，不是你一個人的錯。」我安慰她。

「不行喔！」她頹然的說。

「為什麼不行呢？」我不懂了。

「因為已經吃了太多張罰單了，這最後一張，使我的違規分數大增，要罰我吊銷執照一個月，也就是說這個月不能開車了。」

「除了付罰單以外，有什麼化解的辦法嗎？」我追問。

「可以去上課，不但可以減少一些罰款，也可以減少一些違規的分數。」原來汪阿姐對這些規則這麼清楚。

「那這樣好啦，一個月內，由我開車送你去上課好了！」我又自告奮勇地說。

上課的那天，我先送她去課堂，下車時阿姐手中提了一個極大的袋子，袋中裝滿了東西。

下課時間到了，我去接她的時候，久久不見她出來，停好車走到教室門外走廊，看見一大堆人圍在一起，我擠到外圍，發現她正在忙著做生意，那些一同上交通違規課程的老美們，人人向他購買紙筆、文具、書籍，也有女生向她買胭脂、花粉、香水，最後一位金髮女子，向她買了一個翠綠色的玉鐲。

「買玉鐲的是老師，所以要優惠一點。」她向我解釋。

呀，原來，對汪阿姐來說，此處，他處，處處可以做生意！

Loving mother and her filial son who live in the

Chung Gung Health and Culture villager

Edited by Sandy Chin

When Jade moves into Taoyuan Chung Gung Health and Culture Village trying to live for one month, she feels these "Trial " living is really a great idea! All the seniors live here really like to stay in this most comfortable hideaway! Good size rooms, nice kitchen and clean bathroom, hot and cold running water, heater as well as air-condition, you name it! a variety of food, all of them are delicious, the price is particularly fair, every one of the resident speak Chinese, there is no language barrier between them, you can make as many friends as you wish, you can sit in the cinema watch movies with a lot of people, you can go to travel with many friends, you can also learn painting, flower arrangement, play mahjong , Practice Taijiquan. Staffs are specially considerate and extremely cordial, they have thought of everything for you, best of them all is it's low prices, extremely reasonable compare with the quality of service they have provided. It is really the best place for retirement!

Jade came to Taipei to relax and sightseeing, afterwards she try to look for the free shuttle bus which takes her back to the village, she finds there are some people standing forming a line, looks like they are waiting for the bus, so she walks to the end of the line, after waiting for a while, she becomes   a little uneasy.

"Excuse me,   Is this the line waiting for the bus to take us back to Chung Gung health and culture village?" Jade asks the amiable gentleman who stands in front of her. The man carries quite a few large bags of fruits in his hand.

"The Health and Culture village? Yes, this is the line. Are you new here? " he smiles as he nods his head and answers. It is said that the most beautiful scenery in Taiwan is the people, and seeing him being so affable and polite is an proving evidence.

"There's no obvious bus stop sign, so I am not quite certain." Jade explains to him.

"It's just for people in the Health and Culture village to

form a waiting line it is known by all the villagers that's why the sign is not really needed! Actually the bus should be here any minute now!" He says, and sure enough, there comes a shuttle bus, and the bus stops right in front of them, at the window of the bus there is a sign says " Chung Gung Health and Culture village". When people on the line start to get on the bus one by one, Jade also follow the handsome guy who carried a lot of fruits steps on the bus and sits on one of the priority seat. The guy put his fruit on the luggage self above their head, then also sits down next to Jade.

"I am going to the Health and Culture Village to visit my mother, what about you? Are you going to see… your Mom or Dad?" He asks.

Jade corrugated a bit she can't help but looks at him, and carefully thinks about it, then, she realizes the situation right away, no wonder he has this kind of misunderstanding, because his hair has turned an extremely handsome gray, while her hair is dyed pitch black, before she opens her mouth to tell him that she is doing a "trial living" in the village, not visiting her father nor her mother。

"Give way! let us step aside, Here comes an elderly couple, these are the priority seats, we have to let them sit here!" the handsome guy not only stands up, moves to sits down in the non-priority seat at the back but he also pulled up Jade and have her sits next to him.

"Does your mother like living in the health and culture village?" After the bus started, Jade asks him casually。

"Let me tell you, the problem is the following; my mother likes this village living very much, because there are variety programs she likes, so she does not miss the outside world, However, she is not considering my, that is, her son's position at all!"

"Since your mother is very happy to live here, isn't it a worry free thing for you as a son?" it takes Jade by surprise。

"To tell you the truth, we Chinese think that unfilial piety is the great sin of being a child, for us Chinese, the so-called filial piety is to have all the generations live clustered together, I am now single, my children have all gone to college, living in the school dormitory, no children for my mother to enjoy her grandkids!" Says the filial son helplessly.

"Do you really have to live congested together to be filial?" she is in a great deal of shock. Because she does not know that people in Taiwan would still have this out of date concept。

"Yes, yes, a great and happy family must have old as well as young ones!" someone has joined their conversation. Jade find that the exchange of views between the two of them, is not only heard by the people who sits next to them, but also all the rest of the people, because all the people in the shuttle bus nod their heads in agreement!!

"Yes, if we don't live crowded together, all the mouth will scold the son is an unfilial child!" There is another sigh.

"I am now afraid of the people's laugh, they would say that you have many children, why live in the retirement village, they may think I am not allowed to live with my children." An old white-haired gentleman sitting in the priority seat, too, says in difficult.

"Now, at this modern time, it's still ...." Jade can't believe that she is hearing this.

"Us Taiwanese idea is that the family has to huddles together to generate warmth and affection!" Now, all the people in the bus have heard it, and everyone is talking at once expressing their opinions.

"Now all my aunts take turns to come to my house to point at my nose scolding me, said that I am unfilial, am a bad daughter-in-law, it is so hard to prevent their mouthing!" A middle-aged woman, probably visiting her mother-in-law, says with a being wronged way.

The bus arrives at the destination, There are several villagers; elder ladies and gentlemen standing at the front of the village waiting there, one of them on the crutches is a white hair senior woman.

"My dear son, thanks for coming to visit me! I am assured that I am very happy here!" cried the old lady to the handsome guy who carries a lot of bags full of fruits.

"Mom, this is my new friend, and she's here to visit ..., " the handsome man is about to introduce Jade to his mother, suddenly the cell phone in Jade's bag starts to ring.

"Sorry, I have to receive an international call." Jade apologizes and walks aside to answer her phone.

"Mom, no matter how you slice it, the flight between San Francisco to Taipei will take eleven to fourteen hours, and tell me how can we do it?" That is Jade's son said to Jade on the other end of the telephone.

Jade has no answer to her son.

At this moment, the handsome man came over with a smile and carries one bag of fruit in his hand.

"Hi, I tell you, in order not to say that I am not filial, I have decided to retire next month, complete the retirement process, sell my house in Taipei city, and then I also will moved into the Chung Gung retirement village to live! This bag of fruit is for you and your father, I hope you like it." The handsome guy told his new friends Jade about their mother and son's latest decision.

"It was my son's call from the United States, and he asked me to hurry back to America!" Jade says to the kind and amiable mother and filial son.

養生村的慈母和孝子

一住進桃園龜山鄉老年人住的養生村，如玉就覺得“試住”此村的這個主意真不錯！老年人住在這裡，真是像住進了最舒服的世外桃源！房間好；有廚房、有廁所、冷熱水、冷暖氣，一應俱全，食物種類多，樣樣都美味可口，價錢又特別公道，全部都是中國老人，沒有語言的隔閡，朋友眾多，大家一伙坐在電影院看電影，眾人一同出去旅遊，打伙兒學習畫畫、插花、打麻將、練太極拳。工作人員仔細又親切，樣樣事情都替你想到了，尤其是該村收費非常便宜，實在價廉物美，真的是太理想了！

如玉到台北來散散心，逛完了想要搭乘出來的免費車回村中去，看見有一隊人排隊站在那裡，好像是在等車，就走了過去站在隊伍的最後面，等了一陣子，她又有點不放心。

「請問，要到龜山鄉的養生村是在這裡排隊嗎？」如玉問了一下站在她前面的那位先生，這位先生手中提了大袋小包的水果，看起來很是和靄可親。

「養生村嗎?是喔，是在這裡排隊。妳是新來的嗎?」他笑嘻嘻地點著頭。都說台灣最美的風景就是人，看他這麼和顏悅色，彬彬有禮，就可以證明。

「沒有明顯的站牌，所以不太確定。」如玉解釋道。

「這只是給到養生村的人排隊的，熟人熟地，所以不必站牌，你看車子不是來了嗎？」他一面說，果然前就來了一輛大巴車，車窗上面註得明明白白是龜山鄉養生村的交通車。排隊的人魚貫上車，如玉跟在這位提了很多各種水果的先生身後也上了車，如玉一上車就坐在博愛座上，這位帥哥把水果放在寄物架上，也挨着如玉併排坐了下來。

「我是到養生村去看我年老的母親的，妳呢?是去看媽媽呢還是爸爸?」他問。

如玉楞了一下，不由得抬頭看了他一眼，再仔細想了一下，方才明白了，難怪他要這樣誤會，因為他的頭髮已經有著非常非常帥氣的花白了，而自己的頭髮是用染髮劑染過，所以漆漆黑，正要張口告訴他說是自己本人在養生村親身試住，并非去探望父、母親。

「讓位，讓位，有一對老年夫婦要上車了，這是博愛座，我們必須讓位給他們！」這位帥哥不但自己站起來，還把如玉也扯了起來一同坐到後面的非博愛座上。

「你媽媽喜歡住在養生村嗎？」車子開了之後，如玉隨口問他。

「告訴妳，問題就在這裡，我媽媽非常喜歡養生村，因為裡面節目眾多。搞得樂不思蜀！一點也不替我這做兒子的著想。」

「既然你媽媽住在這裡很高興，那你做這做兒子的不是省心了嗎？」這真是出乎如玉的意料之外。

「不瞞妳說，咱們中國人認為不孝是做兒女的大罪，所謂的孝道就是要數代同堂，我現在是單身，我的兒女們已經上大學，住在宿舍裡面，沒有小孩來給我母親享受含飴弄孫之樂！」這位孝子無奈地說。

「難道真的住在一起才算孝順嗎？」如遇嚇了一跳，沒有想到台灣的人原來仍然有這種想法。

「是喔、是喔，家裡要有老有少才熱鬧嘛！」沒有想到他們兩人的對話，被車中旁邊的人聽到了，不但有人接口說道，而且全車的其他人全部都點頭同意呢。

「是啊，不這樣，眾口紛紜，都罵那做子孫的不孝哩!」另外又有人歎道。

「我現在就是怕人家見笑，說有兒有女，居然要住在老人院，以為我是不容於兒女哩。」一位坐在博愛座上白髮老先生，也很為難地說。

「現在這個時代了，還……。」如玉簡直不敢相信親耳聽見如此說法。

「咱們台灣的觀念，是一家人擠在一起，才算有溫暖，有親情！」現在，全車人都聽見了，大家七嘴八舌，搶著發表意見。

「三姑六婆，七姨八嬸，天天都有人到我家來指著我的鼻子罵我，說我不孝，是個壞媳婦，眾口難防啊！」一位中年婦女，大概也是去探親的，很委屈的告訴大家。

車子到達目的地，只見養生村前站著好幾位老先生、老太太都在等人，其中一位扡了拐扙的白髮老婆婆，見了帥哥由車上下來，歡天喜地地迎了上去。

「我的兒呀，叫你放心，我住在這裡很好!」老太太大声嚷道。

「媽，這位是我新認識的朋友，她也是來探望…。」帥哥正要把如玉介紹給他的老母親，不意如玉皮包中的手機驚天動他地響了起來。

「對不起，我要接一個國際長途電話。」如玉道完歉，走到一邊接電話。

「媽，不管妳試住得怎麼樣，趕快回來罷，由舊金山到台北要飛十一到十四小時，叫我以後如何去看妳呢?」如玉的兒子在電話的那頭對如玉說。

「⋯⋯。」如玉不知如何回答兒子的邀請。

正在此時， 這位帥哥笑嘻嘻地走了過來， 手中提了一袋水果。

「嗨，告訴妳，為了免得親友們說我不孝順，我已經決定下個月退休， 辦完退休手續，把台北房子賣掉之後，自己也搬進老人院來住了!這一袋水果是送給妳和妳爸爸的, 希望妳們喜歡。」帥哥告訴他的新朋友如玉關於他們母子商討後的最新決定。

「剛才是我兒子由美國打來的電話， 他要我趕快回美國呢!」如玉也只得對兩位善良可親的慈母和孝子說道。

母與子

Happy Seafood Eating In Florida

Edited by Adela Karliner(08/dd/2020)

Years ago, we purchased some waterfront marshland in Florida, and we moved there when we retired. While waiting for the workers to build our house and a private dock to store our boat, we worked in the state park as volunteers.

The park's riverside is abundant with large firefish (grunt) and blackfish (mullet). Live shrimp or small fish are excellent bait for the grunt; they are always hungry. Without any effort, each casting would yield plenty of fish. The fresh firefish's flesh is delicate. Poached firefish dipped in soy sauce, with green onions and shredded ginger, is a heavenly food! The mullet does not eat live bait. The Park rangers taught us to cast our nets. Mullets caught by fishnet are big, fresh, and without any bruises. You can grill, smoke, fry, or braise them, and they taste delicious.

Since there are so many of them, the local people came up with a kind of entertainment called "mullet tossing."

The patch's location and size decide the white chalk to paint the flat ground guidelines. The nearest and largest one is $10.00, and the smallest and the furthest marked $150.00. The crowd cheers loudly and claps hands happily when a large bucket of fresh mullet arrive—each fish weighs about two pounds. The player comes up with $3.00, then he or she can choose one fish and toss it. Each time they hit the target, all the bystanders cheer loudly and clap their hands happily once more! Fortunately, they donate the collected money to the Florida Education Fund. Otherwise, what a waste of natural resources! No wonder at one time when the Taiwanese dollar appreciated, the Taiwanese buyers sent people to teach Florida locals to use special knives to break open the fishbelly to harvest the fish eggs, then frozen and shipped directly back to Taiwan. After properly processed, it tastes delicious regardless of how do you cook them; steaming, grilling, frying, or braising.

It is especially magnificent to cut the fresh mullet into sections to make soup; bring a pot of water to a boil, throw in sliced ginger and chopped garlic, some soy sauce, and put the sectioned fresh fish into the boiling water, along with some green scallions. Cover the pot with the lid, then you take off the cap, the steaming aroma hits your nose right away, and it makes your mouth water! In your soup bowl, the clear soup floating with a little bit of fish oil is the world's best! Too bad that we could not share it with our American friends because they do not eat the fish head; they call our favorite soup 'broth, ' and use it as stock to make soup.

One day, park manager Dom carried a giant pot. In it, there were carrots, tomatoes, onions, potatoes, and many snapping turtle cuts. He shared the thick soup with everybody and gave each person a generous bowl. It takes a strong man to cook turtle soup. When fresh, the soup is so delicious! The turtle's skirt is soft and fatty. The meat contains a rare protein-based acid that taste like chicken, deer, cattle, sheep, and pigs-- five kinds of meat, called "gourmet five meat, " Its texture is closer to beef but much more delicious. Unfortunately, after two days, the leftover in the refrigerator had a strong fishy smell!

After building our house, we added a smaller boat to our collection. Now, we were as happy as "the fish in the water!" Sitting in the smaller boat and floating on our backyard canal, we watched the dolphins swimming by our boat.

During the season for bay scalloping, groups of friends would organize into a flotilla in the clear sky, white clouds, and blue waves. The red and white flags on each boat fluttering in the wind were very eye-catching. The bright flags signified that people are surrounding their craft were diving and swimming and that all the passing vessels must be careful not to hurt the divers in the water.

After throwing the anchor into the sea, we tied the scalloping net to our waist and jumped into the water one by one. The scallops opened their double shells in the water grass, spitting out the seawater, then thrust quickly swim forward. When a diver's hand touched the scallop's hard surface, they reached out quickly and grabbed it, then raise their hands with the catch to show the crowd. After cheering and yelling, the team members immediately cleaned their yields on the ship, digging out two small muscles that connected the shells with a knife. It's 40 to 60 pieces per pound. Steamed fresh bay scallop is as tender as tofu, plus it has the umami seafood taste. The outside of the fried scallop turns crispy golden brown, and the inside stays soft; it is a dish that goes well with either rice or alcoholic drinks.

Florida's seashore is also rich in blue crabs. On the surface of the river, floating with the waves, are colorful crab-cage buoys. It is a beautiful water scenery.

Casually place rotten fish, shrimp, or chicken neck in the pen, and throw it into the water the night before. You will fish up a cageful of blue crab the next morning. No matter how we fry, steam, or braise them, the two of us can't possibly eat them all.

"What to do? The crabs stay in the cage will kill each other and eat the other crabs alive!"

"Yes! How about giving them to our neighbors?" A flash of light came to us! When we were very proud of our cleverness, what did we see? We suddenly noticed that our neighbors were carrying a basketful of giant live crabs to visit us with a big smile!

Really!

Now that we are older, although we have given up fishing and crabbing, we still live in the same house, not far from the river. There is a rustic tiki bar selling drinks and fresh seafood. With $12.00, you can buy a box of a dozen giant, fat crabs cooked in the southern Cajun style, with fragrant spice. They claimed that it is a dozen but often deliberately throw in a couple more for adequate measures. It also comes with a box of lettuce, Western-style seafood sauce, plus a box of seafood cream used to dip the crab meat. It is a large bright red box, a great tasting delicacy.

In the 17th century, the French colonized the Acadia, Canada cooking techniques, and their descendants moved to places such as New Orleans in the southern United States, which also profoundly influenced Floridians' cooking. The spices used in this seasoning are small fennel seeds, dried thyme, fennel seeds, bell peppers, hot peppers, dried Oregano, mustard, pepper, onion, garlic and other mixed and ground spices. It is delicious. Boiling fresh crabs, shrimps, and clams in these spices not only gives off a tantalizing aroma, but the red color is also mouth-watering.

Florida is a sanctuary resort. People from the north also called 'snowbirds, ' come south to Florida to avoid the coldness of the northern winter. Every year on Saturday before

Christmas， the Homosassa River holds a light-boat competition.

The Wang family invited friends and family to their backyard to watch the parade. We were among the guests. We busily prepared two boxes of seafood and side dishes and arrived early at their house. Soon， the guests arrived one by one. Some people came with smoked fish， some with prawns， and some others with stone-crab claws. People in the yard were making a fire for barbecuing seafood. Some people started to eat and drink on the balcony while watching the boat teams' colorful bright lights passing by. If the passing one were an old friend's light boat， everybody would immediately put down the food and drink and starts to dance up and down and cheer loudly. The people on the ship would respond by turning the sound of the loudspeaker up. We did the bar-hopping from Wang's to Lin's， then to Zhang's to eat and drink and wait for the lantern-colored river fleet to get the evaluation， desk's score. When the lantern fleet came back through the Lin and Wangs' families' backyard， we jumped up and down again， shouting with raised arms， our hearts all beating faster， and got excited once more!

佛羅里達州水鄉的幸福食緣

多年前先在佛羅里達州水邊早買了地，一退休就搬了過來，趁等候工人們造房、砌私家船塢的那段時間，我們先在佛州公園做義工。

公園邊河中火魚（Grunt）、烏魚（Mullet）一來就是一大群，火魚可以用蝦或魚肉做餌，它們全是餓魚投胎，常常一桿數魚，得來全不費工夫，加以新鮮火魚肉味十分鮮美，白灼後，加醬油、青蔥、薑絲、油淋一下就是一道人間美味！烏魚們不吃活餌，公園管理員教我們撒網，用魚網捕撈的烏魚又大又多又新鮮，用這些魚來燒烤、煙燻、油炸，吃都吃不完，當地人就想出一種娛樂，叫做《拋烏魚》，把泥土地劃上很多格子，按遠近標價，近的 10 元，最遠的 150 元，眾人興高采烈地抬來一大桶兩磅左右的新鮮烏魚，只要出 3 元就可買一條來丟，丟中了大家歡呼鼓掌，幸好收到的錢捐作佛州教育基金，不然真是暴殄天物呀！難怪曾在一度台幣升值的時候，有咱們台灣的收購商派人來教當地人用特製小刀破開肚皮採收烏魚子，冷凍後運回台灣。我們將網來的新鮮烏魚清理後，無論清蒸、紅燒、油煎，怎麼都好吃，尤其是切塊氽湯，將鍋中的清水放入薑絲、蒜瓣、醬油，滾後把魚塊丟入，加上青蔥，蓋上鍋蓋燜一下，吃的時候，碗中浮著點點魚油，真是世間絕味，可惜不能與外國朋友們分享，因為他們不吃魚頭 管我們最愛「喝」的清湯叫做 Broth，是拿來做湯底用的。

一天, 公園管理員唐姆抬了一巨型海鍋, 裡面有胡蘿蔔、番茄、洋蔥、洋芋等各種作料，濃濃稠稠的鱉魚羹來與大家分享，這種大隻切塊的 Snapping turtle，一般都是由力大身壯的男士來清理烹調，他分給每人一大碗，又香又鮮，那真叫好吃！膠質的鱉裙柔軟而肥美，鱉肉並含有一般食物中少有的蛋基酸，具有雞、鹿、牛、羊、豬 5 種肉的美味，故有「美食五味肉」的美稱，口感比較接近牛肉，卻比牛肉鮮美得多，可惜吃不完的在冰箱中放兩天之後，魚腥味就出來！

房子造好，我們又添加了一大一小兩艘漁船，這下，真是「如魚得水」了！開了小船在後院河中釣魚玩水，在落日的餘暉裡觀看海豚游水嬉戲。佛州灣貝開放的時候，我們呼朋喚友組織的採貝船與其他的船隊在青天、白雲、藍浪中行駛，各船上紅白色的旗幟同時迎風招展，非常醒目，每條船的四周都有人在潛水，駛過的船隻一定要小心注意，不要傷害到正在水中的採貝人。

把船錨拋入海中後，我們腰繫採貝網，一一跳下水去，只見一隻隻灣貝開合著它的雙殼 在水草中靠著吐出海水的推力迅速地向前游動，伸手快捷地摸到硬硬的灣貝，高舉手中的捕獲物示眾，在船上的隊員們立刻高聲歡呼。用小刀將連著貝殼二粒小小的肌肉挖出來，每磅 40 個到 60 個不等，新鮮的清蒸灣貝，鮮嫩無比，有豆腐的柔嫩，兼有海產的鮮美，炸成噴香的金黃色美食，外脆內軟，送飯下酒都是上品。這裡也盛產藍蟹，河中的水面上漂浮著點點美麗彩色的捕蟹籠浮標，隨波浮沈，成為一道亮麗的水上風景。

隨便用什麼腐魚爛蝦雞脖子放在捕蟹籠內，前晚丟入水中，次晨撈起一籠藍蟹，兩個人無論油炸、清蒸、紅燒，怎麼也吃不完。

「怎麼辦呢？吃不完的螃蟹留在籠中，自相殘殺，把別的蟹活生生吃掉了！」

你看，如何是好？

「有了！送人如何？」靈光一閃，計上心來！兩人正自慶幸自己腦袋真行之際，只見鄰居笑嘻嘻地帶了一大簍活蟹過來送給我們！

真是！

現在年事已高，雖然放棄釣魚採貝捕蟹，但沿河不遠處的水邊，有一個鄉下的 tiki 酒吧，兼賣下酒新鮮美食，美金 12 元就可以買一打用南方卡真（Cajun）式烹調得又辣又香，隻隻肥大的螃蟹。說是一打，卻常常故意多送二隻，每打加附上一盒生菜、西式海鮮醬、另加一盒用來沾蟹肉的奶油，紅艷艷的一大盆，供人細品慢嚐。

卡真是十七世紀的時候，法國人殖民到加拿大 Acadia 所用的烹調術，其後裔又移居美國南部紐奧良等地，佛州也深受其影響。這種調味料運用的香料計有：小茴香籽、乾百里香、莞荽籽、甜椒、辣椒、乾奧雷根、芥茉、胡椒、洋蔥、蒜等混和研磨成香香辣辣十分惹味，煮出來的螃蟹、鮮蝦及蛤蜊，不但發出誘人的香氣，使人垂涎欲滴，而且看起來油紅光亮，令人食指大動。

此間是避寒勝地，這兩天天氣轉涼，北方怕冷的人們，陸陸續續南下到佛羅里達來報到，聖誕節前的星期六，河中舉行燈船比賽，王家事先邀好親朋好友，到他家後院相聚，也在受邀之列的我們，不等天黑，就忙忙帶了兩盒卡真海鮮及配菜，早早抵達王家，不久，本小鎮的大小老中們，有人帶燻魚，有人帶大蝦，有人帶石蟹腳 Stone crab claws，眾人一面在院子裡生火燒烤海鮮，一面在陽台上邊吃喝邊觀看河中五顏六色爭奇鬥艷的燈船隊伍由家門經過，若駛過的是老朋友的燈船，那就要立刻放下正在享受的吃喝，跳起來手舞足蹈，全體高聲歡呼以示捧場，船上的人也把擴音器聲音提得更高以示響應，等卅多隻燈船一一過去，大家就再度呼朋喚友，又到另外林家去吃香喝辣，等待張燈結綵的遊河船隊由評審台評分完畢，回頭時經過林家後院，再度跳上躍下，振臂高呼，心跳加速，再度胡亂興奮一次！

The Village Elder

Edited by Descartes Li

My husband, who was full of zest every day, would drive the car out of the house in a rush. Then before long, you would see him come home in a hurry. His days were busy, and he 7 usually in high spirits. I feel he could be regarded as Th Village Elder in our lightly populated but sprawling township of Homosassa, Florida. Let me analyze the situation for you.

The first element of being a village elder is of course seniority. He had been retired for many years, and it might be that he wasn't very old at the beginning, but the fact that his hairs on top of his head had turned from peppery to snow white had said it all; He was the eldest of the few Chinese people in town.

Felling that he had traveled all over the world, extremely experienced, plus imbued with modern knowledge of science and learning, he considered himself one notch above of Confucius.

Confucius said that among any three people who traveled together there will be my teacher, but my husband would follow the principle that among any threesome I must be the THE teacher, Even on a guided tour, he would frequently correct or rebut the guide's introductions.

"Not 30, 000! It is more than 335, 000!!" He would say.

I would gently tug at his elbow and wisper:" come, come, as a tourist guide he has to rely on this to earn a living, don't take it too seriously!"

"Knowledge is the truth!"

So, there is no need to compare him with Confucius, we already know whose knowledge of Taoism is greater!

After his retirement, the first big thing was to go sea-fishing with his friends and family. "The ocean is so limitless! The fish waiting to be caught exceeds millions and billions!" he announced. Can you tell me, the situation is or not totally critical!?

In Homosassa, we had two motor boats, a smaller one

docked in our backyard river, while we had a couple of replacement motors, the smaller boat itself remained the same. On account of the increased safety awareness of the village elder, the other boat, which was already the larger and could hold four people, was recently traded in for an 8-person boat. In order to haul that larger boat to the water, we got a bigger car. At home , the village elder purchased the upgraded radio system with a satellite connection for the weather, plus walkie-talkies to connect with friends.

If the weather forecast was clear, then he would initiate a busy and serious series of events. The village elder would coordinate all the participants, buy bait, another to tak care of a picnic lunch, and he himself would rush out to fill his fuel tank.

What about when the weather was bad?

"A 70 percent chance of rain." The satellite radio would report, along with shockingly loud alarm blasts.

Once, after the screaming sounds had lasted for a while, I lifted my head to look out of the window, the rain had been heavy forsome time. Where ws the 30 percent chance of the clear weather?

It could happen in the middle of the night. Suddenly, sound sleeper would be awoken from a sweet and quiet dream by a loud alarm blast!

"Now, please be aware that there is a typhoon level seven warning. It is extremely dangerous, please return all ships to harbor." The radio would announce.

"what has happened?" I rubbed my eyes with fear and anxiety.

The village elder said:" That was a warning for the boats still at the sea, our boats are already hauled into our home made boat-shed, They are safe and very well protected! I have foresaw that this might happen!"

I looked out of the bedroom window and, all was pitch dark. "This radio is so good, if there is any danger at sea, it will broadcast a warning!" He was very proud of his radio.

I put my face in my pillow and could not decide should I laugh or weep.

Sometimes, when it was not raining, but the wind was

oniy blowing make big white caps, the village elde not able to go to fishing. He would stroll about the near by flea market. There he might find a second hand lawn mower, and after the some wise and clever baraining, he might drive the winning plunder back ho e. Then, very happily, he would add oil, and sit on the driver's seat. Although later he might find out that themowing blade had been loose, with a scre had long disappeared. Plus, perhaps all the parts of the machine were held together only by the rust! But he would begin the repair process of this newly acquired machine. For the next few days, he would be so busy that his white hair would be as fluffy as that of Einstein, and his whole face smeared with greasy oil, He had had no time to wipe his sweat nor had he found any time for eating! And soon enough, he triumphantly drove his newly repaired lawnmower out. Now, he had also neglected to water the yard, so that here was now no grass to mow in our two-acre yard.

Next he would drive to the grass-selling company, buy a large amount of grass seeds. He also prepared the yard and fertilized it. After finally sowing the seeds all over the bare soil, he would examine the sky in the day time and at night time, he'd stare at the moon and the stars above. At the same time, he listen for the weather on his radio.

I could not help asking him. "Do you hope it rains or hope it clears up?"

"If it rains I don't have to water my lawn, if it turns out to be a clear weather, then we will go out fishing!" was his answer.

Ah Ha! The village elder had set himself in a fool proof situation!

In addition to being busy at home, he liked to involve himself in the affairs of our Chinese friends in town.

"Mr. Li, here is a letter in English from the county court, would you please take a look at it?" A new immigrant would ask him.

"Well, he court has summoned you to be a witness, and you will have go to court in person." He said after reading the letter.

"Professor Li, the judge speaks English, and I o not

understand English. What should I do?" asked the fellow.

"I'll go with you, okay? Do you have a car?" asked Professor Li.

"No, I do not have a car."

"Okay, tomorrow morning at nine o'clock, I'll pick you up at your house." So, the village elder found himself doing interpreting at law court.

Now according to Florida State law it says that every restaurant must have a qualified food certificate before it can open.

"Dr. Li, the certificate test questions in the state exam are in English, and the answers must also in English. If we can't pass the test, our restaurant will be forced to close!" The owner of the restaurant complained to him.

"What is the test about?"

"The issue of food hygiene." The Chinese restaurateur replied.

"Do you have a book for the person who takes the test to read?"

"Yes, we do." The boss lady quickly went in and handed over a thick book.

"As long as there is ONE certificate for the whole restaurant, it will be OK. It doesn't say the owner must pass the test." Our Princeton PhD , retired biochemistry professor/village elder flips the page of the book. Then, told the boss and his wife.

"All you have to do just to hire a certified person to be a food onsultant and hang the certificate on the wall." The restaurateur knew all this all along.

"Well, then, I'll go and take a test for you." The village elder said with great generosity.

One week later, the restaurant owner drove the village elder, plus their t wives to test site. After about three hours, the village elder came out.

"How was the test?" The three of us asked with great concern. The future of the restaurant was at stake!

"The test was not difficult but it was really lengthy." Our

biochemistry professor answered, wiping the sweat from his forehead.

After two weeks, food safety test certificate was recived by our scholar.

I saw him driving in and out, busy for many days.

"What are you so busy about? People are waiting for your certificate, so it can be affixed to the wall, Otherwise, the state attorney will come to seal up their restaurant!" I said with anxiety.

"I am afraid that they are reluctant to spend money, and wanted to affix the certificate on the wall with some ordinary homemade paste. So, I myself specifically went to the town frame store and ordered a noble, high class and much more appropriate frame. Which was now framed by an elaborate golden frame.

So his designation as "The village elder" and not the "wise man in the village", was particularly fitting, don't you agree?

Professor Li passed away on July 25, 2014 at Crystal River Hospital in Florida, and his friends, Dr. Gu Shun and Dr. Herching Huang, donated ESS Library one, in 2015 at Dahe School in Dingxi City, Gansu Province. for memories.

The Professor Li's survivor Gwen LI recorded this information to show her appreciation and wish the friendship lasts forever.

Library donated for Professor Li Luku, 2015

Principal Zhang Guoqiang e-mail 2844918422@qq.com mobile phone 18139900507

Agents in Gansu Province

Head of ESS Library, Jia Mingwen

村中長者

話说我家老公，每天興致冲冲，開了車出去，不久，又看見他開了車回家，日子过得煞有介事，忙忙碌碌，可以算是我们河漠沙沙鎮老中裡面的"村中長者"了！讓我分析给你聽罷。

長者第一要素，当然是年長，也就是年纪大。他已经退休多年，就算当初退休時，年纪并不怎麽樣，可这些年下來，日月蹉跎得不長也不行了，加以原本花白的头发，现在已经变成是白雪一般的事实，可以用来证明他的年齡的老大，何况蜀中無大将，廖化作先锋，我们這鎮上，就那麽寥寥可數的幾个中國人中，他的年纪的確比较大些呢。

他自認游遍天下，经验丰富，并兼俱现代化的科學知識，所以比孔子更勝一筹。

孔子说过，三人行必有我师，我家老公却時時遵守三人行我必为师的原則，甚至连外出旅行，他都要频频纠正反驳導游的讲解说明。

"哪里是三十万，已经三十三奌五万有余啦！"他会这样说。

"嗳，人家導游不过靠这个赚取一点生活费用，不要太认真罷！"我在旁边轻轻地扯他的手肘。

"知识就是事实嘛！"他还振振有词呢。

所以，把他与孔子比，誰的道行比較高，我们也不必去深究了罷！

退休以後的第一件大事，就是与朋友们出海釣鱼。

"外面大海這麽遥遠遼阔，其中待捕的大小鱼兒何止千萬！"他這樣宣佈，你看情况嚴重不嚴重！？

我们本來就有马達船大小二艘，小船是风大浪猛的天氣裡在自家後院逛逛之用，所以年來只换了一个新马達，一切照舊。近来有鑑於年事日長，为了安全起見，就把原來只能载四人的船，换了一艘新的可坐八人的大船，为了拖大船入水，又换了一辆大車，因怕家中的電話被電脑佔線，另添了两支手机，以便与釣友们随時連絡，這是硬體的升级。軟体方面呢，專报氣象的那台收音机，每逢必要時就先發出各种不同的声音，然後就有電脑控制的播音。

播音宣佈天氣放晴，由他自已替馬達油箱加油，另外誰去買餌，誰带午餐，幾奌钟集合，運籌帷幄，調兵遣将，当然忙得不亦乐乎。

天气不好的时候呢？

"氣象预报，有百分之七十的機会下雨。"收音机裡面宣佈。

但聽雨声嘩啦震耳，我抬頭朝窗外看，大雨明明已经下了很久啦。另外那百分之卅放晴的機会在那裡啊？該時也，家中有線電话及無線手机响声不断，都是在讨論天氣问题。

有时三更半夜，萬籟無声中突然警告的铃声大响，把睡得迷迷糊糊的人，從梦中惊醒，一身冷汗，由床上跳起来。

"现在，各位注意，附近海中有七级风浪，情况十分凶险紧急，大家请将船驶进附近来港湾躲避风浪。" 收音机如此报告。

"怎麽一回事？"我驚疑不定地揉着双眼。

"那是警告在海中行驶口的船隻，我家的船早就拖进院子裡面，未雨绸缪嘛！" 他十分庆幸。

我抬頭看臥室以及窗外，到處乌黑一片，伸手不見五指。

"這台收音机真好，只要海上有警报，它就自動打開，進行报告了！" 他很得意。

我把臉理進枕頭内，哭笑不得。

有時天不下雨，但是大風吹得海中白浪淘天，無法出海，他就施施然到附近的跳蚤市场中，找到一架二手割草机，买後欢天喜地地开回来，加完油後，坐上驾驶位，才知道车下割草的刀刃已经鬆了，螺丝早就不知去向，只靠铁銹把各种零件团结在一齐，连忙跳下来修理，一连数天，忙得头发蓬松上豎得跟爱因斯坦不相上下，满脸油迹乌黑，甚至顾不得擦汗，当然无法吃飯了！

等到把割草机修理好，兴冲冲地开出去，这才发现我家两亩地的院子里面無草可割草。又忙忙乱乱地开车到賣草种的公司，购买了大量的草种回来，满院子低头撒种，终于把种子撒完，後来就頻頻仰头，白天细察天空中的云彩，夜观月亮星斗之外，还要看電視，聽收音机，希望天上下雨，免得自己澆灌草地。

"你到底希望天下雨呢？还是希望放晴？"我不禁问他。

"那还用說，下雨滋潤草种，放晴就去釣鱼。"原来自認是智者的老人家把自己立於不敗之地。

还有，他除了在家中忙碌以外，还要插手河镇别的中国人的事情。

"李老先生，这里有一封由法院寄来的英文信，请您老人家给看看好吗？"一位由国内来的新移民向他请求。

"唔，法院傳你作证，到时你務必亲自到法庭。"他读完信后，对来人这样说道。

"李教授，法官说的英语，我们一点也聽不懂，怎么办呢？"这位同胞请求。

“那我陪你去好啦，你有车吗？”李教授不放心的问。

“没有。”

“明天上午九点，我到你家去接你好了。”他自告奋勇。

我们佛罗里达州的规矩，每一家餐馆都得要一张考试合格的食品証书，才能开业。

“李博士，州立考试的试题都用英文，答案也得用英文，若是不能通过，我们餐馆就要被迫关门啦！”我们常常去的的那家中餐館的老板，忧心幢幢地向我老公诉苦。

“都考些什么呢？” 老先生开口问道。

“食品卫生问题。”中国老板很诚实地回答。

“有书看吗？”

“有。”老板娘飞快地进去拿了一本厚书出来。

“整个餐馆内只要有一张证书就可以了，任何人员有都行，并未规定老板必须有证书。”我家普林斯顿生物化学博士把书翻了一翻，告诉老板夫婦。

“是呀，我们只要聘请一位有证书的人做食品顾问，把他的证书挂在墙上就可以了。”原来老板及老板娘早就知道。

“那这样好了，我去考一个证书，免费做你们餐馆的食品顾问好啦！”他非常慷慨地说。

老板开车，带了我们一行四人到一小時路程以外的鄰鎮，讓他在试場内考了三小時左右才施施然出來。

“考得怎麼樣？”事关餐馆前途啊！

“问题雖多，并不困难。”他一面擦著满頭大汗，一面回答。

后来，州政府果然由邮局信件中寄來了一张食品知识考试合格证书。

一連數天，只见他开车进进出出，忙忙碌碌，煞有介事。

“你忙些什么呢？人家专等你的证书，以便贴在墙上，免得州检察官来查封他们的餐馆呢！”我忍不住问他。

“我怕他们舍不得花钱，用普通漿糊将証書草草贴在墙上，所以特地到鎮上相框館订制了一个高尚、大方、精緻的相框，你说好不好？”他高举鍍金的相框，很得意地问。

所以，他只能算“村中長者”，并不是“村中智者”，不知你可同意？

当然，長者也有他可爱的一面，例如他常聽人說;人老了，味兒會重些，所以他就經常刷牙漱口及洗澡， 到了那裡，身上带著漱口水及肥皂的香氣就跟到那裡，若是有人稱讚他：「好香呀!」他就會露出孩子氣的笑容，在你身邊轉來轉去呢!

李教授不幸於 2014 年 7 月 25 日在佛州水晶河医院仙逝，好友顧洵、黃河清博士夫婦於 2015 年落实在甘肅省的定西市安定區大河學校，捐贈 ESS 圖書室一間，以為記念，未亡人余國英，特錄此文以致感谢之情，并祝友誼長存。

李盧嘏教授捐贈的圖書室，2015

校長　張國強

e-mail 2844918422@qq.com

手機　18139900507

甘肅省的代理人

ESS　圖書室負責人　賈明文

FAMILY WITH SIX TREASURES

Edited by Leah Karliner (05/30/1992)

(01)

The Li family lived in Glen Cove, a middle-class community on the north shore of suburban Long Island, N. Y. White families, all living in ranch houses on one-third of an acre, with Green lawns in the front and decked yards in the back were their neighbors.

Their ranch house was quite large enough when Li's purchased it, but one by one, three years apart, with the arrival of each of their six one-thousand ounces of gold, they needed a little more space. And, as the Li's raised their six daughters, the contractor lifted the roof on their single-level home, providing a second level.

Property values on Long Island increased so much while the Li's six daughters were growing up that their American neighbors, some with one or two thousand ounce gold children of their own (some had ten thousand ounce gold treasures), easily kept pace with the Li's and renovated their ranch-style houses into two-story homes. Externally the Li's house looked just like all of its neighbors. There was only one minor difference: the Li's were the only Chinese people living in the entire neighborhood.

Mrs. Li is a pleasingly plump woman of medium height, with a sweet, good-natured personality.

The Li girls come in a wide variety of shapes and sizes; all are pretty and well-built, with the beauty of flowers, smooth and precious as jade.

And then, there is Professor Li, the solitary red flower amidst ten thousand green leaves. He is a tall, slim gentleman, with gray hair, and wears gold-rimmed glasses befitting his scholarly manner.

Sometimes people ask Professor Li, "Do you have any regrets that you don't have sons?

"If my daughters do not marry foreigners, then, I have no regrets. If they marry Chinese men, their babies will have pure Chinese blood; Grandchildren born of daughters will be virtually same as grandchildren born of sons." Over the years, Professor Li has repeated these words so often that they might as well carved on his forehead.

Their Number One daughter, Amy, comments, "Dad isn't serious, is he, Mom?"

" That's unreasonable. The only guys we know are foreigners," Celia, the Number Three daughter, protests.

" From my point of view, Daddy didn't start early enough educating us on his theory of marriage." Adds daughter Number Two, as she pursed her lips and pushed her glasses up. Her name is Betty.

Throughout, Professor Li has been sitting quietly, reading his paper.

The six Li daughters' names are in alphabetically and sequentially order. The first daughter's name is Amy. The next one is Betty. Celia is third; Diana is fourth. Number Five, the youngest one is Elsa. And the youngest daughter, Number Six, is Fran. Very clear; easy as A, B, C; without any confusion.

Addressing all of her girls, but Betty specifically, their mother says, "didn't Daddy always say you should marry your own kind?" Mrs. Li sticks up for her husband, right or wrong.

"In my opinion," Number Two Daughter says, looking at her Dad out of the corner of her glasses, an impish smile on her face, "What Dad should have done on the day Amy was born was to get two kittens, one white and one yellow. Then Mom and Dad should have taken turns saying that the yellow kitten is better looking than the white kitten, and then when they were grown, the yellow cat is better than the white cat!!

The mother and the six girls all laugh.

" Actually, the yellow cat is better looking than the white cat." Diana tries to stifle the laughter that is welling up inside.

Their father can no longer ignore their teasing. Of course, he doesn't feel that they are funny and says quite seriously, "When God created human beings, he put them in the oven to bake. Those that he took out before they were done became Caucasians. The ones which were burned, but still edible became black. And those which were baked just right, to a golden color, are the yellow people. And, the most perfect of these got to be Chinese."

Teasingly, Mrs. Li states, "Your father is a scientist putting forth his scientific theory. If any of you can develop a more scientific theory, we'll believe yours." She chuckles softly while the six one thousand ounce treasures roll with laughter.

" Daddy is right." is daughter number four, Diana's contribution. "The Chinese Kung Fu fighters are the best looking of all."

Elsa, Daughter number five, says what she knows her other sisters are thinking; "Really?"

Diana, visualizing the handsome Chinese actors who perform Deeds of derring-do in Hollywood's movies, climbing trees and leaping over tall buildings, answers, "Yes, really. Look at the Chinese movie stars. They are all handsome: Bruce Lee, Jackie Chan, and Dean Deng."

"Dean Deng? Is that his real name or a stage name?"

"Dean Deng makes so many movies. Two of his most famous are "Dean Deng, The Bell Sounds" and "Dean Deng on the Church Steeple."

"Are you pulling my leg?"

"Would I lie to you?"

"When it comes to Kung Fu, Of course you would."

(02)

Professor Li doesn't mingle with his Caucasian neighbors. His favorite Chinese proverb is, "Everyone should sweep snow from own front door and not worry about the ice on other roofs."

Being very busy with her family and her housework, Mrs. Li also had little involvement with the neighbors. Her command of English was not subtle enough to gossip easily with the other women: or was there any time for friendly visits to her neighboring homes. Occasionally friends of her number Four, Five, and Six daughters came to the house to do their homework together or to play. But these were teenaged girls between the ages of 11 and 17. Daughter number One Amy is already 26 years old and has been away at Columbia University.

Professor Li teaches at Columbia University. That's why Amy went there. Because her father is a faculty member and her grades were so good, Amy did not have to pay tuition for her under-graduate or graduate studies. Now she has her master's and a license to teach. She come back home to teach English at Glen Cove High School, her Alma mater.

Living at home saves money, and there's always someone there for company. However, there is one big problem; Mom is still dropping broad hints to remind Amy of her age.

"There is a well-known Chinese proverb. 'When a boy reaches that certain age and becomes a man, he should marry and have children bear his name. And, when a girl grows bigger, she should marry someone on whom she can depend.' According to the Chinese way of reckoning age, you are almost 27 years old."

"Marriage is the most important thing in a woman's life. Do you have a 'duwee Shun' yet?"

"Duwee Shun? What's a 'Duwee Shun,' Mom?"

"Duwee Shun means male friend, or boyfriend."

Among all the sisters, number Four daughter has the most incredible understanding of China and the Chinese language. She is a passionate reader of the Kung Fu stories, all about love, hate, politics, and romance among the Kung Fu fighters. She loves them all. And, as a bona fide fan, she watches as many Kung Fu movies as she possibly can.

"How come I never came across that word in my Chinese language class?" Amy wonders.

Amy's hair is black, cut into a short bob, and is very shiny. Her almond-shaped eyes are deep black and beautiful, her lips like tiny sweet cherries set in a small round face. She is an intelligent and attractive young lady.

"Big sister's boyfriend teaches Physics at our school. Mr. Aponte," Elaine, number Five daughter, and Glen Cove High School student, interrupted.

Elaine is a sophomore at the school where her sister teaches. In addition to talking about boys, another favorite topic of conversation among her girlfriends is the single teachers' love life: who is the boyfriend of which available female teacher and which woman is each of the bachelor teacher dating.

Elaine adds, "His full name is Anthony Aponte."

Amy must be in love. When she heard the name Anthony Aponte, a deep crimson blush swept across her round face. A small giggle escaped, and she hides her face. Amy thought quietly for a few minutes; then she raised her head and asked, rhetorically, "Dad is really against us marrying a Caucasian, isn't he, Mom?"

Mrs. Li tried to find a way to answer her daughter, which won't make matters worse, "Your Daddy has often said that he wants Chinese son-in-laws."

Number One Daughter's face became pale. Softly she postulated, "Actually, many Caucasian have very dark straight hair too. And their skin isn't necessarily pale."

Elaine, the Number Five Daughter, frequently interrupts to get attention. This discussion is no exception, "Daddy doesn't want foreigners as sons-in-law because he's afraid they'll laugh at his peculiar way of speaking the English language."

"Maybe he can't keep up with his colleagues and that's why he hates foreigners."

As usual, Mrs. Li has compassion and says, protectively, "Your father teaches bio-chemistry and as long as he is knowledgeable in that science, how well he does or doesn't speak English is truly irrelevant."

Number Three Daughter came into the house during the latter part of this discussion. She's wearing a mini skirt which reveals just about all of her long, shapely legs. The matching shirt is a modified tank top, short enough to lay bare her tiny waist. Celia waited for her Mom to finish speaking to offer her opinion, "Dad's hatred of Caucasians stems from the fact that the Chinese girlfriend he had before he met you, is married to a Caucasian."

Mrs. Li disagreed and said, "You kids shouldn't invent stories like that!"

With a shake of her head, accompanied by the tinkle of her long metallic earrings, Celia defended herself, saying, "Elaine told me about it."

Elaine explained, "Libby, one of the girls in my class is Amer-Asian. One day, she yelled at me. She said 'Your father is a home-wrecker.'"

"What?"

Elaine continued, "Libby said, 'My parents are always fighting and it's your father's fault. My Caucasian Dad says that my Chinese Mom used to be you father's girlfriend. My Mom tells my Dad that if he hadn't lied and told her that your father returned to China to marry a girl his parents selected for him, she'd be married to your father right now! Then my Dad gets very angry and that's why I hate you!'"

"Does Dad really have an ex-girlfriend living in Glen Cove?"

"Well, Mom always told us that a long time ago, after the girl Dad had been going out with got married, he went back to his home in China. Lots of match-makers tried to fix him up. When he saw Mom's picture he said 'OK, that's the one'. He married Mom and brought her to America."

"What's Libby's mother's maiden name?"

"Who knows? I only know that Libby's last name is Smith."

Mrs. Li is stunned by what she's heard. But, for a long time, she doesn't say anything. Her daughters continue their speculations. Finally, she regains her composure and, with somewhat a forced smile, says, "Other people's domestic fights have nothing to do with us. Your father works very hard, spends a lot of hours commuting to New York City; so just don't mention any of this to him. Besides, he's probably forgotten all about the incident."

Then on the theory that the problem of Amy's unacceptable 'Duwee shun' would solve itself if it were left alone, Ms. Li told Amy, "in as much as your Dad is so opposed to any of you marrying a Caucasian, let's not say any thing about Mr. Alpo...the Physics teacher."

"Don't mention it! Is it like an ostrich burying its head in the sand?" The second daughter Betty asked.

(03)

One weekend, when things were too quiet to suit her, Celia became the instigator and said to her father, "Dad, you really and truly think it's not a good idea for any of us to have a Caucasian boyfriend?"

"Of course I don't."

"But, you've never had a white son-in-law. How do you know that they're no good?" Celia continues, winking surreptitiously at her sisters.

In utmost seriousness, Professor Li announces, "At my age, I've eaten more salt than you've eaten rice and..."

In chorus, the girls interrupt him to complete the Chinese

proverb he has spoken to them repeatedly throughout the years, "...you've eaten more salt than we have eaten rice and you've crossed more bridges than there are roads upon which we have walked."

All six one thousand ounces of gold laughed uproariously.

Fran, the youngest one, told her Dad, "Daddy loves salt!"

"There are as many Chinese proverbs as there are stars in the sky," commented Betty.

"Well, there are 5,000 years of Chinese history. If each of those years, the mother made up one proverb and the father discover a second, in 5,000 years, we'd have 10,000 sayings. If each proverb were a star, then, It certainly enough to take care of our sky.

About a year after that, Amy picked a time when her Dad wasn't home to tell her mother, "Anthony and I have been looking at houses every spare minute. Today we saw a small one that we really like. You know, Mom, with the prices of houses going up the way they are, the houses we're looking at now are much smaller than the ones we looked at when we first started looking. If we keep looking much longer, then, we will not be able even afford a bath room!"

On a sunny Saturday, shortly after that, when the birds are chirping and the blossoms blooming, Anthony Aponte drove up in a two-door Accord. Amy had invited meet her folks. She was waiting outside to meet him. He has straight hair, almost black eyes, and his skin is practically the same shade as Amy's. Standing there, they certainly look like a couple who belong together.

Amy wished fervently that, for once, her father would just let things develop naturally.

Among the six one thousand ounces of gold, Betty wore glasses because she was slightly near-sighted. Number Four Daughter's eyewear had thick glass, the kind the kids call bottle glass. So, with these six girls, one mother and the one physics teacher who has just arrived, there were 20 "eyes," every single one of them staring fixedly at Professor Li as he appears in the front doorway of the house. They wanted to see if he had any reaction to Anthony.

When he first saw Anthony Aponte at the end of the driveway, because Professor Li saw Oriental-looking dark hair ad familiar skin tones, there was no change in his demeanor. However, as Anthony came closer, Li realized that the man with his arm Amy was not Chinese. His face got very red, and he said rudely, "Do you know why I don't like Caucasians? Do you?"

They all knew he was not expecting an answer. The young man was very embarrassed. He smiled deferentially and, having rather quickly removed his arm from Amy's waist, nervously rubbed his hands.

Father Li continued, "Caucasians are barbarians. No tradition. They do not know filial duty. They know nothing about sibling love. Filial duty means you listen to elders and do what they say. And elders are the man's father, the man's mother, the girl's father, the girl's mother. All   the in-laws and out-laws, both sides. Sibling love means you must love all your brothers and sisters. Our Amy has five sisters. Do you love and care for them all?"

The more Professor Li talked, the angrier he got, and everything he said was in English. Who said that Professor Li couldn't handle English? After all, language is a communications tool. Didn't he make his point clear enough? Didn't everyone understand his opinions?

The men stood face to face; four eyes stared at two eyes.

Mrs. Li could almost feel the electricity between the two, and it was getting tenser by the moment. Quickly she motioned to Betty and Diana, saying, "The Dim Sum is ready!"

Number Two and Number Four are Professor's pets. Each of the girls grabbed one of his arms, and together, they maneuvered him into the dining room, with everyone else following. There was not much joy and laughter as they sat and to eat lunch and drink tea. Their afternoon repast ended, but the tension did not.

Amy was smart enough to know that her father would not do as American fathers of the bride do; pay for the wedding extravaganza. That would be like asking a tiger to give away his skin. Forget all about it! And getting a dowry to bring into the Marriage as was Chinese custom was just as unlikely.

Shortly after that, Amy and Anthony found a house. The next time Professor Li was away from home, working in the City, Amy and Mrs. Li went shopping for a pattern and some white fabric. In the bridal outfit they created, petite Amy looked like a Chinese doll. Anthony tried on tuxedos and rented one that made him look handsome and sexy.

The following Wednesday, a bright and beautiful day, when Professor Li was safely at work in Manhattan, Amy and Anthony dressed in their wedding garb, went to the Glen Cove City Hall. Crowed into the office at City Hall were the groom's widowed mother and on Amy's side the mother-of-the-bride and five younger sisters. Anthony's mother was as quiet and polite as he. On the Li side, there was much chattering and tittering. The only time there was silence was during the three minutes it took the Mayor of Glen Cove to recite the wedding vow's words.

That very afternoon, on returning the tuxedo to the rental shop, Mr. and Mrs. Anthony Aponte took up residence in their house, so new that the paint wasn't dry. Their two-day honeymoon was spent in old clothes, fixing up their love nest.

Glen Cove High School is around the corner from the Li's house. At first, the married daughter stopped home often to have lunch with her mother and shared all her complaints about married life with her Mom. Usually, her husband dropped her off, but, occasionally, he also ate lunch in the Li kitchen. Before the end of a year, he dropped off two people at the Li's house: Amy and a chubby, bubbly, adorable, mixed breed baby girl.

(04)

Mrs. Li realized that her husband has not reconciled himself to the Marriage of his Number One Daughter and is, in fact, very unhappy about it. Their Number Two Daughter has been very much on her mind because Betty will soon be graduating from Columbia University. Then they'll be facing the same issue.

One Saturday, when she arrived home from the supermarket, Mrs. Li didn't follow her usual procedure of putting away the groceries. She sat down next to her husband, who was in the living room, reading the newspaper.

"Today, while I was shopping, I met another Mrs. Li."

"In Chinese, the name Li is the most popular last name, Meeting a Mrs. Li is not noteworthy to be excited about," said Mr. Li, without an iota of interest evident in his voice. He continued to read.

"The Mrs. Li I met is a very hospitable woman. Do you know why she is so hospitable?"

"Of course. They have too much rice and so much money that they don't know what to spend it on," Professor Li pretended that he had no ears and made it very obvious that he did not want to engage in conversation with his wife.

Last night, he watched the game on television; but he liked to read about the game's finer points in the sports section the next day.

Mrs. Li was not about to give up. She was as excited as an explorer who discovered a new continent. She experienced some conflict because she knew that she should take the groceries in. She shouted, "who wants to help me by putting away the groceries?" Almost immediately, three beautiful young girls rushed down the stairs to help their Mom with the groceries.

Mrs. Li sat on the couch next to her husband, lowered her voice, and said mysteriously, "Mrs. Hospitality Li says that she too has lots of daughters and that in America it is very difficult to find that many Chinese sons-in-law. Therefore it is up to the parents to help. So, they invite lots of the young, single male Chinese to their house. They are trying to give their daughters every opportunity to find suitable husband."

When Professor Li heard these words, his face registered. He slapped his leg with the palm of his hand and exclaimed, "Now I understand totally. It's no wonder we were never on the guest list when they had a party. I thought they did not invite us because they are new riche and that scholarly people intimidated them. It is our beautiful daughters that intimidate them. They know their girls couldn't stand such competition!"

Professor Li's mood has changed totally. Now, he is all ears. He wants to know everything about Mrs. Hospitality Li's strategy.

Daughter number Two, Betty, lived in the student dorm on the Columbia University Campus. She was engrossing with experiments, thesis preparation, and applying for admission to medical schools, with little opportunity to go home to Glen Cove.

Suddenly, Professor Li took great interest in Betty's activities. He picked her up at the dorm to take her home for the weekends. But all Betty did at home was staying in her room to study.

Betty's parents were studying too. They were concentrating very hard on the guestlist Professor Li had compiled on all Chinese graduate students attending Columbia University. Names, photographs, curriculum vitae on each student. Professor Li had been even more careful and precise in collecting the students' data to invite to his house than he'd ever been with his Bio-Chemistry experiments.

Now he and his wife are seated at the kitchen table perusing the assembled material. Professor Li handed his wife the folder he has just examined, saying, "We must invite him. This one's father is the Dean of Law School at the National Taiwan University."

Mrs. Li looked at the photo from another folder and told her husband, "This one is very handsome. We must definitely invite him."

"Yes, we must invite him because he is on Dean's list again," countered Professor Li.

"Healthy looking."

"Influential family."

"Father owns a big company."

Gradually, they finished preparing the guest list.

Meanwhile, Betty, who had grown tired of studying in her room, decided that she needed to walk around and stretch her legs. Wearing blue jeans and a matching jean vest, she came downstairs. Her legs were long and thin, and her overall appearance was more that of a high fashion model than a med student. Her dark shiny hair almost reached her waist; she hasn't had time to have it cut. Behind her eyeglasses, Betty's eyes were long and slanted, with dark eyebrows. Her lips were a beautiful shade of red, very shapely and reminiscent of a delicious fruit that grew in the water in China.

Betty noticed that her parents were so deep in discussion that they didn't even realize she was in the kitchen. Curious to know what was engrossing them, she moved closer.

Mrs. Li was planning the party. "Buffet style. I think we should serve a buffet. Should we have wine?"

"Should we?" asked Professor Li.

When Betty heard that her parents talked about serving

alcoholic beverages, she interrupted, "My roommate's cousin needs money. He knows how to serve drinks. He could be a bartender at your party and earn the money." Betty seemed to be very interested in helping her friend's cousin earn money.

(05)

Mrs. Li smiled lovingly at Betty and said, "silly child. We're inviting guests who can help themselves to what we offer them. We don't need a bartender."

"Wait a minute." Professor Li asked, "Can the cousin of your friend drive a car?"

"He has a beat-up old Buick; but he's mechanically talented and whenever the car breaks down he knows exactly what to do to get it going again. He's really very talented," Betty's enthusiastic description made the impecunious fellow sound like a hero.

Hearing Betty's comments, Professor Li said to his wife, "Let's hire him as a helper. All of the Chinese fellows live in New York City. None of them has a car. If he can drive the guests back and forth, serve the wine and help me move some of the heavy things, he will be a good helper. I am not so young any more."

Betty hopped from one foot to the other in her happiness at their decision. She kissed each parent, saying, "that's so great! He gets paid $60 for a whole day. Now he'll make some money and he won't have to worry about the money any more!" Then, she flew up the stairs to call with the good news. Her intention of walking around to stretch her legs has flown, too, all the way to Timbuktu.

Professor Li thought about how different the customs in America were from that of the Chinese. He could not remember a time in his life when he had hugged or kissed his parents. It just was not done. This 'hugging and kissing' is one American custom pleasing to Professor Li and Mrs. Li. The two of them smiled and looked at each other for a while. Then the Li's continued their guest list discussions.

Every time she saw or talked to Betty in the coming days, Mrs. Li reminded her that she must come home to play hostess for their party. To Mrs. Li's surprise, not only did Betty agree immediately, but she also brought the roommate's cousin home with her on Friday night, much earlier than the Li's had expected.

The roommate's cousin was a white American with bright blue eyes and very blonde hair. He was not so bad looking. But his clothes were on the shabby side, jeans washed out to white, and both jacket and jeans were quite threadbare.

"Oh well. He is only here for a day," Professor Li said to himself. He wrinkled his forehead with displeasure but said nothing as he took the young man Betty has brought to the basement to move around the heavy furniture in preparation for the party.

"He must have been fed much bread and butter," the Professor speculated, recalling that the strength of Americans had always attributed to the fact that they ate plenty of these staples at an early age. "He really is strong and healthy," Professor Li thought that he had already gotten his money's worth from the young man, just in the energy that saved his old bones.

At three o'clock, Professor Li said, "here is the guest list with the addresses," as he handed it to the Caucasian.

Betty, who has been hovering nearby, says, "his name is Bill."

Bill took the list and went to his car. He opened the Glove compartment to take out a street map of New York City. Betty climbed into the front seat with Bill, and together, they studied the plan to lay out the most efficient itinerary. Shortly after, a bombastic noise emanated from the car, along with heavy black smoke from the tailpipe. Bill started the engine. Slowly, very slowly, the car moved forward accompanied by a cacophony of clanking.

Professor Li saw the car moving and quickly ran after the car, yelling, "Stop! Stop!"

Betty poked her head, ponytail swinging merrily back and forth, out of the window and asked, "what's the matter, Dad?"

In Chinese, Professor Li said, "If the car breaks down on the way to New York and everyone turns to see what is the matter, it is a big scene but of little significance to you. If the car breaks down with guests in the car and delays them, it puts your entire future in jeopardy and is of great significance."

Question marks marching across her forehead. Betty pushed her glasses up and shouted, "What?" She has studied

Chinese for more than two years, getting "A's" all the time, but she didn't understand the meaning of the Chinese words her father had just spoken.

Her father solved the problem even though Betty still did not understand that there was a problem by saying, "Tell him to drive my car." He gave his car key to Bill.

(06)

When the Professor saw that Betty was again getting into the car with Bill, he said, "You don't have to go along with him, let him go by himself. You change into nice clothes and get yourself ready."

Obediently, Betty got out of the car.

Professor Li added, "and since you will not need your glasses to read tonight, leaves them in your room."

It took Bill two roundtrips, Glen Cove to New York, to pick up all of the guests, during which time Betty got ready. She was usually too busy studying to pay attention to her natural good looks. Tonight, our number two treasure dressed very simply in an apple green dress, with matching flats and a matching silk ribbon to tie her ponytail. She wore no makeup, but without glasses, her pretty eyes were all the adornment she needed.

"How is it without your glasses?" Mrs. Li asked Betty with concern.

"Without glasses everyone looks friendlier and sweeter," the Number Two Miss Li answered. She was only slightly near-sighted, and not wearing glasses didn't affect her too much.

While Betty and her mother talked, the Professor watched his fourth daughter, Diana, come into the room, decked out in a brand new Kung Fu outfit and without her glasses.

"And you are doing what here?" Professor Li questioned Diana.

Diana has read that long ago in China. Eligible daughters participated in a Kung Fu contest with any eligible male guests who could best the daughter would be the son-in-law. Not realizing that finding a husband for Betty was the object of tonight's mission, Diana thought that one of the Chinese males visiting their house might be an expert at Kung Fu. So she told her father, "I just learned a brand new set of Kung Fu move that I saw in Dean Deng's latest video tape. We can have a contest tonight, to see if any of the guests are as good as I am."

"How come you are not wearing glasses?'

"I am a good and obedient daughter." Diana blinked her eyes. "Didn't you say we shouldn't wear our glasses?"

Observing Diana further, Professor Li asked with great concern, "Is something wrong with the food? Why do you smell each dish as it is brought to the buffet table?"

Mrs. Li clarified, "Diana isn't smelling the food. She is just a near-sighted young lady, not wearing her glasses, who wants to know what we're serving."

There is a Chinese saying that no one knows a daughter as well as her mother. It certainly is true in this case.

"Oh," said the Professor. He would like to explain to Diana why she should wear her glasses tonight, but he has to take care of his guests now and can't take the time. As Professor Li looked around the room at all of these promising young men, one of whom could be an essential part of Betty's future, the father felt much happiness in his heart. Then he noticed Betty, and his joy momentarily ceased.

"Betty. That Caucasian worker should be able to take care of himself. You don't have to keep on feeding him. Come here and I'll introduce you to Mr. Sung. His father is a Professor at the college of Science and has a reputation for being one of the best teachers at the school."

Betty moved gracefully, her steps as light as the wind blowing the willow as she walked over to shake Mr. Sung's hand.

Professor Li continued, "and here is Mr. Fong, Betty, He comes to us from the Peoples University in Fu Chow. Bring him a glass of wine."

Promptly Betty obeyed, getting the glass of wine from Bill.

As Betty did that, the Professor said to his Number Four Daughter, "What is wrong with Diana? Why are you sniffing the guests' faces?" Immediately, he answered his question, "Of course, she is not wearing her glasses."

Oh well. Everyone has already observed Diana's peculiar behavior. Nothing anyone says will change that.

He turned away from Diana to look for Betty. Again she

was feeding Bill. He hurried to Betty and said, "Look, Bill is the hired hand. You already brought him soap from the kitchen. You do not have to keep bringing him food," the father steered Betty towards another young Chinese man who was standing in the corner at the opposite end of the room and said "Here is Mr. Chang. He is from Qingdao."

(07)

Since the beginning of the party, all of the Professor's eyes had been working overtime. The first eye was to see what peculiar action Diana would take next. Two eyes needed to watch all of the young men, to make sure that they all noticed his beautiful, brilliant Betty. And the fourth eye was doing double duty tracking Betty so that he could keep reminding her that she was not the guardian of Bill's stomach.

All of this activity exhausted him.

Suddenly he realized someone was missing.

"Where is Diana?" he asked himself, apprehensively. "What can she be up to now?" Finally, Number Four Daughter has stopped sniffing guests and gone back to her room, probably for further Kung Fu studies.

Then, the Professor caught sight of Betty flitting around the room like a pretty green butterfly, going from one Chinese student to another, each with a great future before him. The venerable Professor's heart was enlightened.

Father Li was so happy and laughed so much that his mouth was rarely closed. And, he couldn't help drinking a few glasses of wine more than his limit. No problem. Bill had the car keys and could take all of the guests back to New York City. The Professor didn't wait for his guests to leave. He just went to his bed and before he knew it, was sound asleep.

The next morning when he woke, he found the other side of the bed was unoccupied. Putting on his robe, he went downstairs to the kitchen. There he saw his three youngest daughters getting ready to eat breakfast. Fran was pouring milk into the bowls of cereal.

He asked, "where is your mother?"

In her sweet childish voice of a twelve-year-old, Fran replied, "Mom is in the basement. When we finish eating breakfast we'll go and help her. Do you want corn flakes, Dad?" She continued, reaching for another bowl.

"No. No. Fran. No. Please, I don't want any," he was horrified at the thought of this form of American torture. "When you chew them, they make such a crunchy noise in your head; it hurts! I couldn't stand all that crunchiness!"

Not even stopping to get a cup of coffee, he hurried downstairs where his wife was cleaning up the basement. He started to help, asking, "Where is Betty?"

"Betty said that she had to get back to school to study; so, when that worker came to get his money, she got a lift back to the city with him."

"When did they go?" the Professor wanted to know.

Mrs. Li thought for a moment. "It was about 3 o'clock in the morning. I was kind of sleepy and didn't look at my watch."

No wonder. Last night he dreamt that Betty was wearing a red Chinese bridal gown and a red veil covering her face. She was marrying the best of all of those young Chinese men who were here last night. In the middle of the ceremony, some Chinese Firecrackers exploded. That must have been the Caucasian's car clanging as Bill drove his antique vehicle away.

Professor Li was somewhat disappointed. As he stacked the dirty dishes, he talked to his wife. "In your opinion, which of the young men did Betty like the most?"

Mrs. Li thought very carefully and answered diplomatically. "Betty was polite to each of them. It's really difficult to tell if she favored anyone."

Now the Professor was disheartened. He worked quietly for a while. Almost to himself, he said, "did we waste energy, time and money?"

Mrs. Li didn't want her husband to be even more disappointed, so she said, "well. All of the young people including Betty live in New York City. Now, they know each other. They are acquainted and future developments are possible. If they don't know each other, nothing at all can happen; so it's most important for them to be introduced. Now, who knows what will happen when they see each other."

Professor Li thought his wife made some valid points because the first step is essential. Subsequent developments would be in the hands of Fate and luck.

(08)

After the party, Professor Li had a new mission in his life. Whenever he went to work in the City, he also checked up on Betty's activities. He wanted to find out what was going on so he could discuss it with Mrs. Li. But all of his checking up brought no good news. Many of the Chinese guests were interested in Betty. After all, they would be marrying an American citizen, and as the spouse of an American, they would have no trouble staying in America.

However, when any of them asked Betty for a date, she was always too busy. No movies. No parties. Oh sure, when she saw one of them in the cafeteria, she would sit down with them to eat her sandwich. But, never twice with the same person. Everyone knew that Betty would be graduating soon. But, no one knew where she did her studying. There was no sign that she was dating any male Chinese students, nor was there any evidence that she had fallen in love with one of them.

Li and his wife can do nothing. These are modern times. No longer can match-making be used to arrange for a girl's future. Shun Chin is out of the question. That was the custom still practiced in some parts of China, where a family with an eligible son planned to eat dinner with a family with an eligible daughter. At the dinner, the offspring would be the ones who decided if they wanted to meet again.

There were still five unmarried daughters. The parents were getting as jumpy as ants walking on a hot wok.

One day soon after the party, Professor Li came home earlier than expected. Betty, who had come home that morning, heard his car and ran outside to greet him in the driveway.

"Good news. I have good news to tell you," the number two daughter Betty shouted before her father could get out of his car.

"Are you going to get married? Ah. You found a Chinese husband? That's why you are so happy!" Professor Li guessed.

Betty didn't hear her father and kept waving an envelope in front of her father's face, saying, "good news, Dad. The Medical School at the University of California, San Francisco is interested in me. They want me to come for an interview."

"Oh, Very good, Interview means they want to talk to you," Professor Li knew that this interview would be a formality. His daughter was so pretty, so intelligent, and would speak so well that no interviewer would refuse to accept her.

"When do you go?" he asked, a hint of sentimentality in his voice and face. He knew that Betty's leaving this time would be like letting a bird out of a cage. The chance of her coming back to them was much less than it was when she left to go to school in New York City.

"Tomorrow," Betty answered, her eyes shining with excitement.

"Then you had better get yourself ready," the father was sad at the thought of letting her go. Then he thought how painful it must be for his wife, as a mother, having to watch her babies grow up and leave home. He went inside the house and comforted his wife. After all, they had been holding hands for half a lifetime.

Just before dawn the next morning, Professor Li got up and looked out the window. There was Betty, carrying a big backpack, standing next to a large piece of luggage. Waiting with her were Mrs. Li and Diana, both keeping Betty company.

Professor Li threw on some clothes, ran down the stairs and outside. He wanted to say good-bye to Betty.

"So early," he said.

"Oh, we have to start early. If his car breaks down, we'll need extra time to fix it. Normally, it takes about two and a half days to drive straight through. But, we're going to allow half of our two-week school break to get out there. And then, he wants to show me around the area so I can get acquainted with everything."

From a distance, a clanging noise accompanied Betty's words. The noise hit Professor Li's ears, and It was a familiar sound. An old shabby broken-down Buick pulled up in front of the house. Professor Li's eyes have seen this configuration before. The car door opened, and a handsome, very healthy-looking, and lively young man jumped out.

Professor Li's brain searched through its memories. Yes. Yes. There stored the look of the faded blue pants and threadbare top! If this blonde, blue-eyed person was not Bill, the worker from the party, who else could it be?

Was it a dream? The Professor bit his lips. That hurt; it was not a dream. He blinked his eyes vigorously, again and again.

No, it was no mirage. It was him!

Betty was only too happy to explain, "Oh Dad. It's Bill. You remember him. You paid him $60. He just graduated from the School of Medicine at Columbia. Now he's going to be an intern at the University of California's Medical School in San Francisco. If I'm accepted there, then we'll be able to be together at last."

Professor Li saw bright points of light swimming in front of his eyes. It was apparent that he was feeling faint. Betty ran over to prop him up; Diana ran into the house for the cold water her mother told her to get. Bill rushed back to his car and returned almost immediately, pulling a stethoscope from a brand new black leather bag.

The first rays of morning sun peeked through the leafy branches of the trees flanking the Li house. Professor Li opened his eyes. Up close, very close, there were two crystal-clear blue eyes reflecting concern and confidence. White fingers, attached to a white hand on an arm with a rolled-up sleeve, held his wrist. As the ray of light touched the pale arm, each string of the blonde hair accentuated.

Professor Li glanced away and met Mrs. Li's eyes, which filled with anxiety. When did those fishtails appear at the outer corners of her eyes? How long had those threads of white been in her hair?

(09)

Precisely as the Professor expected, several medical schools accepted Betty. As expected, Betty selected the Med School at the University of California in San Francisco, UCSF. With mixed feelings, Mr. and Mrs. Li went to Betty's graduation ceremony, accompanied by six daughters and one grand-daughter of diverse origins. Fortunately, Bill was working, so nothing worse marred the celebration than Mrs. Li's tears falling and Professor Li's tremendous sighs.

The day after graduation, Betty and Bill went to the City Hall in Glen Cove to complete the marriage license application. Every member of the family closed ranks, as tight as the bands on a wooden barrel, to keep Professor Li from finding out. It was a conspiracy of silence, designed to keep things quiet on the home front.

Betty had always been very special to Professor Li. When she was little, he would take her fishing and play ball. He treated her like a son he never had. Visions of Betty working in his Lab, following in his footsteps, had always been part of his dreams. Now, his fantasies would be burst like a child's soap bubbles.

It wasn't supposed to be this way in America. But this was almost like it was with the traditional Chinese families who threw away the water in the bucket when their daughters got married because their daughters would never be back, literally becoming a part of their husbands' families.

Both Li's parents were sad and somewhat depressed that another daughter was gone from home. Fortunately, their oldest daughter Amy brought her female baby to visit them quite often.

After Betty left, the Professor's schedule returned to normal. Weekdays he traveled back and forth, Glen Cover to New York to Glen Cove. It was very tiring. Consequently, he liked to sleep late on the weekends until the sun was at least three poles high.

This Saturday, the sun crept through a tiny crack in the tightly-drawn drapes.

Persistently, the rays sought his eyes, determined to disturb his dreams. From the next room, the sounds of throbbing music came to beat on his eardrums. He had no choice. He had to get up, albeit slowly.

Sticking his head into the next room, he saw Number Five Daughter, Elaine, and her red-haired, freckle-faced friend Elsa lying side by side on the rug. The television and the stereo were both blasting, each delivering the highest number of decibels possible. The girls were listening intently, obviously enjoying what they were hearing.

"Hey! Hey! Hey!"

From Daughter Number Four's room, one could hear that the abrupt, explosive sounds. "Hai! Hai! Hai!" Diana has decided that not only will she read all of the Kung Fu novels, watch all of their movies and videotapes, but she will also rehearse the Kung Fu routines using real knives and swords.

One can hear her reciting "In the most advanced stages of Kung Fu. Penetration of the defenses by your enemy's knife and sword is impossible. Hey! Hey! Hey!"

Professor Li stumbled towards the stairs and, using both hands to support himself on the banister. He walked down ever so slowly. His neck was stiff. He held his head awkwardly, as he was afraid that his brain would burst open.

He made it to the kitchen. There was Fran, sitting, reading. What was she reading? A trashy paperback novel, in English, no less. At the far end of the kitchen, Mrs. Li and her Number One Daughter were chatting. Amy's face flushed with anger, and obviously, her mother was trying to comfort her.

"Must be complaints about her Caucasian mother-in-law again," Professor Li thought, bored at the idea. From his wife, he learned that Amy doesn't get along that well with Anthony's mother. But, the Professor doesn't want to become involved with details as small as sesame seeds. Frequently he is tempted to tell Amy, "Look. I told you not to marry a foreigner."

(10)

The only one who was aware that Professor Li came downstairs was Amy's little girl baby, who was sure she had found a wonderful treasure when she spotted her grandfather. "Gung Gung! Bah Bah! Hug Hug!" She wanted to be picked up and held by her grandfather.

He complied, kissing her adorable smooth babyface. Holding her closely, he inhaled her milky-sweet aroma. The baby was thrilled to be kissed and sniffed by her "Gung Gung." She giggled and gurgled her approval.

"Too bad this cute little one is of mixed heritage," her grandpa thought with regret.

At this time, he heard the noise of a motorcycle approaching closer and closer.

"Vroom! Vroom! Vroom!" Professor Li peered out of the kitchen window just in time to see a shiny new motorcycle stopped in his driveway.

A young man garbed in black, black jacket, tight black pants bounded off the bike. His hair was spiked and colored bright red and apple green. It must spray with glue. His original hair color might not be visible, but there was no mistaking that this was a Caucasian. Even more peculiar, a long earring dangled from one ear. When he jumped off the bike, the earring swung back and forth wildly.

The man turned and, with two hands, helped the girl who was seated on the back, off the motorcycle. She was also wearing a black leather outfit, jacket, and tight pants that matched his. The two kissed a passionate joining of feverish lips. The mate to his earring swayed merrily from one of her ears. There was a stud in her other ear. The one pair of earrings these two wore reflected the bright May sunshine, shimmering and radiating light.

Both were wearing black leather gloves, with the fingers cut off. Two pairs of cut-off fingered gloves, one with ten white fingers and the other ten yellow fingers. All 20 of them poking through their gloves. They looked conspicuous, very conspicuous. Professor Li was scrutinizing this pair. Aloud, his voice full of curiosity, he said to himself, "Why are these two strangers making an exhibition of themselves on my driveway? Let them go to their own driveway for their hunky punky."

Suddenly his youngest daughter let out an ear-piercing shriek. She flung her spoon from the right hand to the floor.

"Thunk." "Thunk." sounded the book as it landed next to the spoon. Fran was only a flash of fast-moving flesh and clothing as she raced outside, past her father's eyes. However, the words of her shrill scream lingered in the kitchen long after she was gone.

"Chuck. Chuck Carlson!"

Somehow, Mrs. Li and Amy slipped past Professor Li without seeing them and followed Fran outside.

They were also very excited.

Everyone gathered in a semi-circle around the young Caucasian. Even Diana, Elaine, and Fran managed to get past the Professor unseen. Each was holding a pen and paper, all jabbing and shouting a jumble of meaningless words.

"Autograph."

"Chuck. Sign here. With love."

"Autograph."

The bodies entwined at the end of his driveway have been interrupted, forced apart. Quickly, the Caucasian signed his name.

"For Diana, Love Chuck Carlson."

"To Elaine from Chuck."

"Good luck Fran. Chuck Carlson."

Doors opened up and down the street. Females in all shapes and sizes came from down the block, around the corner, every house seemed. All were converging on the Professor's driveway. All wanted a Chuck Carlson autograph. They pulled at his clothes, pushed the motorcycle, tugged at his spiky hair. The young man was getting visibly upset. The shrill screams of young girls' voices continued, adding to the commotion. He glanced around apprehensively, looking for a way out.

The entire scene was too much for him. Run for his life! He pushed his way through the swarming, buzzing girls, climbing onto his motorcycle, and jumped up to start it.

"Vroom! Vroom! Vroom!" Before anyone realized it, Chuck Carlson's motorcycle roared out of the development, leaving behind only a cloud of black smoke and a gaggle of yelling and screaming girls in Professor Li's driveway. The white young man ran like a dog without a home and disappeared like a fish, just escaped from a fishing net.

The girls were yelling and carrying on for a very long period. Finally, the group dispersed itself, a few at a time going towards their homes, still excited, talking non-stop about their recent encounter.

Now, let's see. There was still one young Asian girl standing at the end of the driveway. She stood there wholly preoccupied, holding onto her guitar. It was Celia, the Li's Number Three Daughter.

Suddenly, she felt her father's eyes staring at her fiercely, his face dark with anger. She was so startled that she ran into the house and up to her room. The Number Three daughter slammed shut her bedroom door and locked herself inside.

Celia lay on her bed. She looked around the room at the famed singer's life-sized posters in many different poses and hair colors, which decorated her walls. Her thoughts turned to Chuck Carlson and his loving kisses.

(11)

Downstairs, Mrs. Li saw her husband slumped on the couch, so she hurried into the kitchen to get him a cup of tea. Silently, automatically, he took the cup and blew slowly on the floating tea leaves.

Quietly Mrs. Li started to talk. "Celia may not be on the

same level as her sisters, academically. You know, in the United States, the music field offers a good future to talented people. And, musically, she is extremely talented."

Her husband still said nothing.

She also sat without talking for quite a while. Then "Remember when we prepared that list of names of the Chinese graduate students? Do you still have it?"

He nodded.

Very carefully, without haste, Mrs. Li continued. "Now that summer vacation is starting, we must insist that Celia leave the dorm and live at home. We will have to watch her more closely and not permit her as much freedom as we have." Mrs. Li paused to observe her husband. "Once more, let's invite all of the Chinese students to our house; but this time, no Caucasian bartender to mix drinks!"

When Celia found out that she was under house arrest, she became agitated. She cried bitterly, refused to eat, and in general caused great turmoil in the Li household. Attending a party to be matched for a husband is the last straw.

Celia refused to cooperate until her mother diplomatically put a new light on the entire situation.

"It's not the way you think. This party is like a showcase for you and interesting oriental entertainment for these Chinese students who are so far away from home," her mother explained.

This idea intrigued Celia. "Well, I'll have to practice my Chinese songs. And, if I sing in Chinese, the entire atmosphere will have to be Chinese. There must be books which illustrate Chinese dance steps and offer other material on Chinese music," She became quite excited and then very involved.

There were tapes to find, songs to locate, books to peruse if she were to stage a professional show. Celia asked her mother and her Number One sister to make her a Chinese costume. By the time it was finished and ready to try on, she could hardly wait to look at herself in the mirror.

When the Professor and his wife observed that Celia had thrown herself into this project, our Number Three daughter Celia calmed down.

"When is the party?" Celia's sisters wanted to know. "She's been practicing day and night; our eardrums are broken and our nerves are frayed."

This time Professor Li dedicated his old body entirely to getting the guests to the party. He brought two car-loads of Chinese graduate students to Glen Cove.

Mr. Chow, one of the graduate students, borrowed a brand new Cadillac from his friend's butler. Mr. Li's second trip from the City, Mr. Chow followed Professor Li with the big borrowed car carrying a third load of Chinese graduate students. The Li's guest list has grown somewhat from its original size.

Both cars arrived at Li's house, where the door was standing wide open, and Mrs. Li waited to greet the last group of guests with a broad smile.

Would you think it peculiar if each of your guests had a funny look on his face as he got out of the car and looked at your house?

Guess what?

Elaine, their Number Five daughter, and her red-headed friend Elsa were standing outside the house. Both were wearing bright yellow sweaters, each with two great prominent red Chinese characters sewn to the sweater's front. The symbols on Elaine's flat chest translated to "Extremely Delicious." On Elsa's well-developed bosom were characters which in English meant "Very Nutritious."

Would you be embarrassed if these were your family?

Professor Li called Mr. Chow aside and asked, "Would you please act as host for a few minutes? As you see, we have a small emergency to take care of."

The Li's herded Elaine and Elsa into the kitchen. Professor Li was too embarrassed to say the words, which would eliminate the girls' puzzled look. "Elaine, please have Diana come downstairs. She can read Chinese and will enlighten you by translating these four Chinese characters. No. Better go upstairs and have Diana explain. And, for Buddha's sake, change out of those sweaters. I never want to see you wearing them again."

Before they can leave, Diana enters the kitchen wearing a stunning new Kung Fu outfit, with a real sword in hand.

"Here I am. And, Celia is not the only one who practiced regularly," Diana tells her shocked parents. "I also practiced my Kung Fu routines with great diligence," implying that she wanted to join Celia's showcase.

"Oh Diana. This time our guests have been invited for Celia. They're not here for you," low flustered Mrs. Li told her Number Four Daughter.

" Oh. Ohhh! You're going to have a party for each of your remaining one thousand ounce of gold." Finally, Diana understood.

Their Number Five Daughter, who had been listening, protested, "We don't want any of your old Chinese graduate students. We want to invite our own friends as guests."

She spoke with the vehemence that only a disgruntled 15-year-old girl could muster.

"Elaine's absolutely right. If you throw a party for me, you'll have to select only avid Kung Fu fans," Diana made what she thought was a most reasonable request.

The Li's had no time to argue with either of them at this point. Guests were waiting downstairs, and Professor Li and his wife were anxious to join them. Because their parents had said nothing, based on the girls' faces' content expressions, it was clear that the daughters thought their Mom and Dad had agreed to their requests.

On the other hand, Mr. and Mrs. Li's faces looked as if something occurred, which they may regret for a long time. But, there was no time to worry now. There were priorities, and when the eyebrows are burning, one must extinguish that fire first.

Professor and Mrs. Li went downstairs to be with their guests that Mr. Chow has been holding the fort very diligently. He has been a great help to his host and hostess. Everyone seemed to be enjoying himself; the food consumed happily, and conversations appeared to be flowing along with the food.

(12)

The highlight of the evening began after everyone had eaten. Chelsea played Chinese melodies on her guitar. She was sometimes accompanying herself as she sang, sometimes letting the music's mournful sounds speak for themselves. Our Number Three daughter danced her solos. And then, she invited everyone to sing with her, to accompany her by beating the tempo on a nearby table or bench. It was evident that everyone enjoyed participating.

Celia was wearing a Classical Chinese dress and, based on a style she saw in the movies, a classical Chinese hairdo. Her sweet singing voice, graceful dancing steps, and organizing talent have worked in synergy to draw all of the guests into the activities. Indeed, no one was bored. Especially not Mr. Chow, who additionally kept himself busy by working the Camcorder throughout Chelsea's performance, and he made sure that he was focused on the third Miss Li the entire time.

When Celia finished, applause thundered through the Li's finished basement. Everyone was cheering and whistling. Mr. Chow poured a Coke for Celia and said to her, "You are terrific. So talented. So clever. Celia, you are really great."

Celia raised her head and felt as if she had met an old friend from her homeland. Her smile was brilliant, like a flower in full bloom. "Chow! How did you get on my parents' list?" She tilted her elegantly coifed head. "Oh yes. Of course, you're in the graduate study program at the Department of History. My parents do love their Chinese graduate students."

"I haven't seen you in ages, Celia. Where have you been keeping yourself?" Chow wanted to know.

Lowering her voice, Chelsea told him, "what can I tell you! My parents have been keeping me home nights so I haven't been able to get into the City. I feel like I'm in prison. If I hadn't been able to practice my singing and dancing, I would have gone nuts."

It is evident that Celia knew Chow, and they know each other exceptionally well from their conversation. Professor Li had no choice but to take a closer look at Chow.

He saw a tall, thin Oriental with a classical haircut, and generally very neat and clean. The Chow wore a pair of black-framed glasses. Atop his dark gray trousers was a brand new navy blue Chinese style jacket. He looked incredibly handsome. Professor Li said to himself, "I'm sure women find him sexy, with those dark eyes twinkling behind his glasses. He must make women feel...."

Seeing that her father's eyes were staring at Chow, Celia smiled and said, "Dad, his name is Chow. He's a graduate student and very talented. He wrote the best-selling biography "The charismatic Choc Chow Chin." Each of the three volumes has been reprinted several times. His second biography,

"About Chang Chug Chui," was so popular that it was a sell out the first day it reached the bookstores. It's been back-ordered since then. And his next book is already in the works."

Professor Li was a bit embarrassed at having been caught scrutinizing Chow. However, when he realized that Celia was both friendly and enthusiastic, he smiled and told his wife to cut a piece of cake for Chow. And he specifically brought a cup of chrysanthemum tea to this handsome young man.

Now that they finished the song, so was the party. It was deep into the night. So Chow would cram half of the students into his car and take them to drop at the entrance of the subway, then; Chow would come back for the rest of the guests.

Both Li's parents were quite pleased with the way the party had gone. And next morning, both let out a sigh of relief. After much bowing and thanking, the leftover young men got into the car with Chow. Mrs. Li was surprised to see Celia sitting next to Chow. Mr. Li just beamed happily. Her husband looked so happy that Mrs. Li decided to say nothing.

The following week, Professor spent quite a bit of time finding out more about Chow. Not only was Chow a talented and brilliant scholar, but his family background was very distinguished too. His parental grandfather was one of the Emperor's most trusted counselors; his father was the commerce minister. And an uncle on the mother's side was the chairman of the province from which Chow came. No one could ask for better credentials.

When Mrs. Li saw how happy her other half was, she decided to make him even more comfortable cooking his favorite meal: juicy stewed pork butt, with tender green baby pea pod shoots, firry hot peppers, and shredded bean curd. With a smile, the family members picked up their chopsticks.

Somebody was missing from the table; Professor looked puzzled and then said, "Where's Celia?"

Mrs. Li, who just sat down at the table, said, "She went to New York city. In fact, she just called to say that she won't be back in time for dinner and that we should eat without her." Professor Li's face darkened. So did the mood at the table. Dinner was eaten silently, with no conversation.

For a change, Professor Li started to pay attention to Celia's whereabouts. 24 hours. 48 and 72 hours. And still, Celia had not come home. Without comment, Professor Li rummaged around in the desk until he found the guest list. He copied down Chow's address.

Very early in the morning of the fourth day, Professor Li, with Chow's address and a street map of New York City in hand, drove into Manhattan. He found Park Avenue and the number indicated on the paper. Now, he parked his Japanese Camry in front of the expensive, luxury Park Avenue building, which is Chow's address.

A uniformed young Black man asked the Professor, "For whom are you looking?" The uniform jacket is white; the trousers are red, with a narrow white stripe running down the outside of each leg. Everything pressed with precision.

Professor Li felt that he had every right; in fact, he had a fatherly duty to be here. Righteously, he answered, "I'm looking for apartment 305. Mr. Chow's residence."

The black man nodded, gesturing for Li to get out of the car. As soon as Professor Li got out, the uniformed man lowered himself into the car and drove away. He stood there, wondering what to do next.

Finally, he said to himself, "Well, since I'm here and the car is gone, I may as well go in."

Inside the building, another uniformed man, this time a Caucasian, stopped him.

"I am looking for Mr. Chow. Apartment 305," Professor Li said. The man escorted Li to the elevator and pushed the button marked "3". Before the Professor got into the elevator, he noticed how plush and thick the carpet was. And, the crystal chandeliers hanging from the high ceilings in the hallways were big and shiny. The paneling in the elevator was dark and expensive looking.   Everything spoke of luxury and elegance.

The elevator stopped noiselessly on the third floor. Professor Li got out and, observing the instructions on the elegant sign, walked right to the Apartment 305. He pushed the doorbell. Just as he was about to try it again, a man dressed like a butler opened the apartment door.

Before the butler could say a word, Professor said rudely, "I'm looking for Chow. Does he live here?"

The butler answered most politely, "Mr. Chow is a guest of my boss while he's writing my boss's biography and therefore, indeed is living here."

"Where is he? I want to see him," insisted Professor Li.

(13)

Still respectful, the butler explained, "I'm sorry; Mr. Chow is not at home right now. Please give me your name, address and phone number and I'll give them to Mr. Chow."

"When will he be back?" Professor Li asked very loudly.

The louder Mr. Li got, the more politely the butler spoke. "He usually comes back quite late. He's rarely here for lunch. Most of the time Mr. Chow is at the library."

Raising his voice to its loudest, the Professor demanded to know, "What library? I'm losing my patience. I want to see him immediately." He is shouting by the time he finishes talking.

"What's the matter?" the loud words awakened the young Caucasian man who came out of a room to the left of the entryway.

This white guy must be the butler's employer, was wearing a pair of very sort red paisley pajama shorts, with nothing covering his bare muscular top. His arms wrapped around a beautiful Chinese girl. The beauty was wearing a red paisley pajama top, revealing legs that look even longer because they were bare.

If this wasn't his Number Three Daughter Celia, then who could it be?

And wasn't the young man with her that unisex looking motorcycle character who had disrupted the entire neighborhood with his presence recently? The fact that his hair was now silver and blue didn't disguise him at all.

Professor Li was furious. So furious that he wanted to teach Chuck a lesson and punish Celia. Before Professor could do more than raise his hand, two bodyguards, one black, one Caucasian, appeared from nowhere. One held the Professor's left arm; the other held his right arm. At this point, Celia realized that the angry man was her father. Quickly she ran from Chuck's side towards her father, ready to protect him.

Simultaneously Chuck also realized who it was and told the bodyguards, "Stop. It's Celia's father."

Turning to the butler, who has been watching the proceedings quite calmly, Chuck instructed him, a bit tongue in cheek, "Chester, please set up the tea service and prepare some Oolong tea for Professor Li. He is a scholar and must be treated with respect."

Turning to Professor Li, Chuck said, "Celia is 21 years old, Professor Li. You're treating her s if she were still sixteen."

Of course, Professor Li didn't want to drink Chuck's tea even though he was very thirsty. Nor did he want to hear Chuck's words. And he didn't even want to look at Celia. The Professor stormed out of the room and onto the elevator.

(14)

The next thing he remembered was pulling into his driveway.

How did he get his car from the garage? Through which roads did he driving home? And, how did he get to Glen Cove, to his driveway? Li doesn't remember any of the details.

He walked into the house and entered the living room, found it in total darkness. As he walked towards the windows to pull open the heavy drapes, he realized someone was sitting on the sofa. And he asked himself the questions that asked at the end of the "chapter" in the classical literature performed in the Chinese tea house, which meant to pique the interest of the tea drinkers so they would return the next day: "who is this person? And, why is the person sitting in the dark?"

The answer came fast and in tearful tones. "Dad, I want to divorce Anthony," Amy had been waiting in the living room for her father's return. He pulled open the drapes, and the minute she saw her father, Amy cried even louder. She sounded pathetic.

"Wait a minute. Wait a minute! Chinese philosophy states 'if you marry the chicken, you follow the chicken; if you marry a dog, you follow a dog. what crime has Anthony committed that you think you have to get a divorce? Don't cry, Amy. Let's talk about it. I'll help you the best I can." Professor Li was getting very nervous.

His wife knew how to handle these things. Where was she anyway?

Amy started to recite her litany, "Dad, Anthony doesn't love me any more. He loves his mother," she cried like a pale blossom dripping with rain. She could hardly catch her breath.

Professor Li patted Amy on the shoulder and said,

"What's the matter?" with much concern.

"My Caucasian mother-in-law comes to my house to see her grandchild. She hugs her and kisses her. Old people carry germs and bacteria that cause sickness."

That's Crime #1.

"Then, she says she wants a grandson, too. Pure sex discrimination!"

That's Crime #2.

Amy spoke with more significant agitation as she continued to enumerate her mother-in-law's crimes. "She says that the baby is cute and good-looking and that her eyes are just like little almonds. Racial discrimination, without a doubt."

Crime #3.

"She comes to our house every day. Definitely invasion of privacy."

That's Crime #4.

Mentally, the Professor was scratching his head, wondering, "How could that be a crime when it is Chinese tradition that the daughter-in-law should live with the husband's family and became one of her mother-in-law's families."

Amy continued, oblivious to her father's reactions. "I told her that she needs more rest and relaxation and that it wasn't necessary to come visit us every day. She started to cry, with absolutely no good reason."

That's crime #5.

"And today she called saying that she can't balance her checkbook and would I tell Anthony to come over with his little calculator to get it straightened out for her."

That's crime #N

Trying to change the subject, the grandfather asked, "Where is the baby?"

In America, everyone has the impression that Asian women are all tender and sweet and that Chinese daughters-in-law are all obedient and respectful. And, it has been his daughter who was enumerating so many complaints about her mother-in-law. What could he say?

"Anthony was fighting with me and he scared the baby with his loud voice. She started to cry. That's very bad for her mental health. So, Mom took her to get some ice cream. Scaring a little baby like that."

That's crime #N+1.

"Bang. Bam. Clunk." Upstairs, something seemed to have come crashing to the floor.

Again they heard "Bang. Bam. Clunk." Something must have broken in Diana's room.

The conversation between Amy and her father came to an abrupt halt. They listened to determine where the noise originated, and quickly, both ran up the stairs, Professor Li a little slower than his daughter.

Diana's room was in chaos. The lamp tipped onto the bed. Four legs pointing at the ceiling was the chair. Kung Fu tapes and books were all over the floor, virtually wall to wall.

(15)

Crawling around in the middle of the mess was their number four treasure Diana, her hands were feeling and fumbling over every small object on the floor. She squinted her eyes to look at a small item she had found.

"Have you lost something?" Amy wanted to know.

"What are you looking for?" asked her father.

"The sandbag hit me so hard that both of my contact lenses popped out," Diana answered. Desperation tinged her voice.

And the sandbag was still swinging back and forth in the middle of the room.

Worried about the cost of replacing the contact lenses, Professor Li asked, "Do you know how to look for those two tiny little lenses?"

At that moment, they heard the noise of the garage door opening and a car pulling into the garage. All three of them rushed downstairs.

Mrs. Li had come back. Amy went to the garage door to greet her mother and her baby. Relief flooded over Professor Li. His rescuer had arrived, and a big load lifted from his should. He decided to escape to his study. He needed peace so he could think about everything that had occurred.

There were quite a few pieces of mail on his desk. He

sighed. Because of his involvement with Celia, he hasn't even taken the time to check the mail. Picking up one of the letters, he stopped abruptly, his brain bombarded by the words on the paper. He got so dizzy he almost couldn't see the comments.

Then they came into focus. "…Frances is pregnant and wants to have an abortion. However, her physical condition is such that there could be complications and therefore, we feel it necessary to notify you…"

Professor Li wondered why they had to notify him about an abortion. Fran was a very healthy child. What medical condition? Fran's pretty, childish features swam in front of his eyes.

"How could this be? Fran was only 12."

Slowly, he re-read the letter. He asked, "Why do thy keep using the name Frances? My daughter's name is Fran, not Frances." He looked again. "Aha. The letter says Dear Mr. Lu", not "Dear Mr. Li."

As if the lightning jolted him, our Professor jumped off the chair and thrust his head into the wastebasket. He must find the envelope.

"Where is it?" He dumped all of the papers in the basket onto the floor and looked at them one by one.

"Here it is! Of course! It's addressed to Mr. Lu at 7 Apple Tree Lane, not…"

See how careless this American mailman was — the Li's live at 7 Ivey Lane. And Lu and Li may be close in the Roman alphabet, but they are very far apart in Chinese.

When did the other Chinese people move around the corner from us?" he wondered. Why hadn't his wife mentioned the Lu family on Apple Tree Lane to him? He started to laugh.

"Thank goodness that she didn't mention this kind of Chinese people to me," he laughed louder. Our scholar Li was glad that he didn't know the Lu's. His laughter got louder and louder. So loud that Amy and Mrs. Li, who were still at the garage door, heard it. And, it sounded so weird that they had to see what was going on.

Quickly, Amy went to the study. Mrs. Li, the baby in her arms, followed. When Amy opened the study door, she saw her father, eyes bloodshot from lack of sleep for the past three nights, and his mouth opened wide, laughing hysterically. He laughed so hard and looked so peculiar that the baby started to cry, hiding her face against her grandmother's body, saying, "Gung, gung. Gung, gung."

The upstairs daughters heard loud laughter and the baby's crying coming simultaneously from their father's study and wanted to know what was going on too. They descended the stairs, and when they got to their father's study, huddled around his desk. Diana peered at her Dad through her thick glasses. Elaine and her red-haired friend were, obediently, not wearing their red shirts.

Instead, they were wearing matching green sweatshirts, each with two Chinese Characters across the front. Elaine's flat chest characters translated to "Low Price" While Elsa's bulging bosom displayed the characters representing "Hot stuff."

Elaine tried to console her father. She walked over to him and said, "I'll always be an obedient daughter. I promise, I'll never marry a Caucasian man. I'll be a Lesbian."

(End of full Text)

Tianjin China: Award of Senior Literary Collection

家有六仟金

（01）

　　李家住在紐約近郊長島格林可夫鎮的中等白人區，家家戶戶一式的三分之一一畝的院子，前有草坪，後有庭院。買來的時候是平房。後來因為李家小的千金一個一個出世，大的小姐也一天一天長大起來，所以請人來把房頂託了上去，加蓋一層二樓。後來長島地皮房價漸漲，老美鄰居們也紛紛把平房改成樓房，所以現在李家的房子的外觀與鄰居還是大同小異，只不過附近好像仍然只有他們這一戶中國人家。

　　李教授高高瘦瘦，灰白的頭髮，戴了副近視眼鏡，一副紳士學者的派頭。李太太中等高度，微胖的身材，十分溫和賢慧。

　　李家有六位環肥燕瘦的女兒，個個如花似玉，一位比一位漂亮可愛。在這群以李太太為首的鶯鶯燕燕里，李教授可以算是萬綠叢中一點紅了。

　　有人問他，沒有兒子，覺不覺得遺憾？

　　"只要我的女兒不嫁給外國人就行，嫁了中國女婿，小孩們流的是純粹中國人的血，外孫也就跟孫子差不多了。"多年來，他都是千篇一律如此笑嘻嘻地回答，答多了，連自己都深信不疑，早就忘了老夫婦倆人前半生為了生兒子而努力，曾經失望過六次的記錄了。

　　"媽媽，爸爸不會認真吧？"老大愛華問。

　　"這簡直不可能嘛，我們每天接觸的都是爸認為的外國人啦！"老三彩華嘰起她的小嘴。

　　"我看，爸爸的家庭教育沒有做好！"老二佩華抿著嘴，推了推眼鏡，斜睨著正在看報紙的李教授。李家六千金以英文 A、B、C 字母來排行的，老大叫 Amy 愛華，老二叫 Bessie 佩華，老三 Celia 彩華，然後 Diana 黛華，Elaine 依華及 Fran 范華，非常清楚，不容易混淆。

　　"教育怎麼做得不好？爸爸不是常他說要你們嫁中國丈夫嗎？"李太太有理無理，向來是護著她丈夫的。

　　"依我說，應該從大姐出生開始，家裡早就養好黃、白、黑三隻小貓。爸爸媽媽天天輪流給我們姐妹耳提面命，黃貓最好看，白貓不好看，黑貓醜死了。"老二佩華一面說，一面似笑非笑地瞧著他的父親。六位千金，連同媽媽都笑了起來。

　　"其實，黃貓是比白貓、黑貓都好看的！"四小姐黛華強忍著笑說說道。

「當初上帝造人，是放在太陽爐中烤出來，沒烤好就拿出來投胎的，就變成了白人，那烤焦了的，無可奈何也只能讓他投胎，成了黑人，那烤得正好的，就是黃人，而最好的黃人才輪到投胎變成中國人。」李教授看家人取笑他，他自己一點都不笑，正色說道。

「你們的老爸是科學家，看看誰能夠編出更有科學根據的道理來，我們就洗耳恭聽。」李太太說這話時發出輕柔的笑聲，六位千金小姐更是笑得前俯後仰。

「爸爸說的極是。所有的武俠，以中國俠客最為英俊。」老四黛華道。

「真的？」

「當然是真的。你看那些有名的武俠明星，哪一個不英俊瀟灑？諸如李小龍、成龍、鄧叮....。」四小姐黛華說。

「什麼凳叮？是真名還是藝名？」

「鄧叮演了那麼多的電影，其中最有名的是'塔上鄧叮'以及'鄧叮鐘響'。」

「他是中國人嘛？」

「長得像極了中國人，所以才那麼迷人呀！」

「是真是假？」

「我會騙你們嗎？」

「當然會！」

(02)

李教授守著他中國人第一的金玉名言，不大與白人鄰居來往，有道是：「各人自掃門前雪，莫管他人瓦上霜。」

李太太因為家中事多，只是送女兒上學，學彈吉他，跳舞等等與鄰居輪流開車有些交往，平常很少與老美太太串門子，就算女兒們偶然有朋友來玩或者做功課什麼的，也多限於四，五、六小姐寥寥可數的十幾歲的小姑娘們。

這是怎麼一回事呢？

大小姐愛華，今年已經有二十六歲。爸爸在哥倫比亞大學教書，她就在哥大讀書、住宿，成績又好，大學以及研究所全都免學費。現在學位、教師執照都有了，又回到格林可夫母校高中來教數學，反而住在家中了。

住在家中又熱鬧，又省錢，可是有個大壞處，就是母親大人有意無意，老是要提醒她的年齡！

"中國有句至理名言：男大當婚，女大當嫁，按中國演算法，你都快要二十七歲了呀！""女子終身大事，要緊啦，有什麼好'對象'嗎？"…，等等。

"媽媽，什麼叫做'對象'？"這六位千金，都是在美國長大的，知道的中國字不少，可是不知道的更多。

"'對象'就是男朋友。"老四黛華在六姐妹中間算作中國通。她最愛看武俠小說，所知道的中國，全是由這些書裡得來，各種江湖恩怨，豪情俠義，她都十分嚮往，在她心目中的中國，到處都是飛來高去的英雄好漢，就像中國武俠電影裡面那樣。

"我在中文課中，怎麼沒有讀到這兩個字呢？"大女兒愛華問道。愛華有短短的頭髮，烏光柔順，東方式迷人的杏眼，黑白分明，圓圓的臉蛋上有個櫻桃小口，又聰明，又秀氣。

"大姐的男朋友就是物理老師艾安地啦！"正在讀高中的老五依華插嘴說道。依華在姐姐教書的學校讀書，女同學之間，除了討論自己中意的男生之外，生活的中心就是關心女教師的男朋友，男教師的女朋友。

大女兒一定在熱戀中，被妹妹這麼一提，圓圓的臉蛋上一片紅暈。她把頭低下，吃吃笑了一陣，突然想到什麼似的，問道："媽媽，爸爸很反對我們嫁給外國人，是不是？"

媽媽想了一下，大概是想如何措詞。她緩緩地說："你爸爸是常說，一定要中國女婿。"

大女兒臉上蒼白了好一陣子，她才輕輕地說："其實很多白人的頭髮比中國人還黑還直，皮膚也不怎麼的白。"

老五依華，年齡小，在一群姐妹中總愛插嘴，他就趁機插嘴說："阿爸不肯要外國女婿，其實是怕他自己的英文怪腔怪調的，惹人笑話罷咧。"

"說不定是洋同事排擠他，他受夠了洋人的氣。"

"你們的爸爸教的是生物化學，學問好就行，英文好不好不打緊啦。"李太太是個忠厚的婦人，忙忙地替丈夫辯護。

"媽媽，什麼叫做'打緊'？"老五依華問道。

這時，老三彩華剛好由外面進來，把大家這一篇討論全聽見了。

"爸爸恨外國白人，是有原因的，因為他以前有過中國女朋友，後來那個中國女朋友被一個白人男子搶去了。"彩華說。她今天穿了一件極短的裙子，把白白的玉腿全部露在外面，上身穿一件同色同樣向像裙一樣的上衣，不但把細細的纖腰露出來，普藥連粉嫩的肚皮都露出來了一小截。

「小孩子，不要胡說！」李太太不以為然地說道。

「媽，這不是胡說，是老五依華說的。」老三彩華搖晃著她的到大耳環。

「在學校里，那個混血的莉莉，常常罵我，說他爸爸媽媽常常吵架，都是我們家爸爸害的哩！」老五依華委屈的辯道。

「什麼？」

「的莉莉的爸爸是個白人，大概她那中國媽媽，原來是我們爸爸的女朋友，她的中國媽媽常常罵她的白人爸爸把她騙了，要不早就嫁給我們爸爸了，莉莉的爸爸氣不過，兩人常常吵架，莉莉因此恨死我了。」

「媽媽，爸爸真的以前有過女朋友嗎？」

「對呀，媽媽不是常常說爸爸當年回國，人家替他做媒，他一概不要，只看了媽媽的像片，連相親都免了，兩人就結婚呢。結婚後，爸爸就把媽媽帶到美國來了。」

「那你同學的媽媽原來姓什麼？」老四黛華追問。

「誰知她媽媽原來娘家姓什麼，做小姐時叫什麼名字，只知道莉莉現在姓司密斯。」老五依華說。

李太太被這突如其來的消息震著了，自己發了半天呆。女兒們七嘴八舌，妳一句她一句沒完沒了，過了好一陣子，媽媽才恢復神志，強笑著鎮定自己，說道：「人家爸爸媽媽吵架的事，你們不要沒大沒小地去管它，爸爸每天開車那麼遠，工作那麼辛苦，不許在爸爸面前提起，自找沒趣，那些事你爸爸可能早就忘記了。」

「爸爸既然反對外國人，我們就不提好了。」李太太又和顏悅色地對大女兒愛華說。

「不提這事，不就跟鴕鳥把頭埋進沙里一樣嗎？」老二佩華問道。

(03)

老三有的時候問爸爸：「爸爸，你真的不喜歡我們有白人男朋友嗎？」

「當然。」

「你又沒有過白人女婿，怎麼知道不好呢？」

李教授就會正色宣佈：「依我這一把年紀，吃過的鹽，…。」

「對啦，爸爸走過很多橋，爸爸吃了很多鹽。」六位千金一致哄笑起來。

「爸爸愛吃鹽！」麼小姐范華笑得最大聲。

「中國的諺語，跟天上的星星一樣多。」二小姐佩華批評道。「中國有五千年文化，每年編出一句諺語的話，五千年下來，也就布滿了天空了。」

匆匆，又過了一年多，愛華乘爸不在家，又對媽媽說：「他近來與我天天去看房子，今天看中了一棟很小的舊房子。現在房價漲得這麼厲害，我們看了兩三年房子，愈看愈小，再不買房子，就更買不起了。」

醜媳婦究竟要公婆，洋女婿也免不了要見面，在一個鳥語花香的星期六，大女兒把洋同事帶了回家。

這位艾安地先生，開了一部日本豐田的二門小轎車，果然直髮黑眼，膚色與東方人相近，文質彬彬，規規矩矩，斯斯文文，與嬌小秀麗的愛華站在一起，正是郎才女貌，一對璧人到。愛華抱著一線希望，望爸爸能網開一面，打個馬虎。

二個男人面對面站著，六隻眼睛，眼對眼互相瞪著。

李太太看見氣氛越來越僵，趕快叫二小姐珮華、四小姐黛華跟她進去，面授機宜。

「點心熱好了，可以吃了。」二、四兩位小姐仗著平時受爸爸的寵愛，一邊一個，挾持著老爸，大家一哄到了餐廳，這一頓下午茶，吃得不歡而散。

愛華是個聰明女兒，在這個關頭，指望李教授按美國規矩，由女方出錢來個大事鋪張的婚禮，無異與虎謀皮，免提了，要造中國習俗，由女方提供大量豐盛的嫁妝，更是休想啦。

愛華與安地終於選中了一棟舊房子。

乘教授不在家，愛華與媽媽上街選樣子，挑衣料，自己做了一件短的新娘白衣，嬌小的愛華穿起來，真像一個東方娃娃新娘，安地租了一套新郎禮服，試穿了一下，真是風度翩翩，一表人才。

在一個風和日麗的星期三，李教授在城裡上班，兩人穿了準備好的禮服，在柊林可夫市政府公證結婚。男方只有一個個安安靜靜的寡母參加。女方以李太太為首，六個姐妹，吱吱喳喳又說又笑，共同擠在一部車內去到市政府，聽一個市府職員念了幾句結婚公文，大家整整安靜了三分鐘。

當天兩人先把新郎禮服送回出租公司，馬上就搬進那油漆未干的新居•第二天，兩人就捲起衣袖，開始整理兩人的愛巢。

反正學校在娘家附近，大女兒三天兩頭往家裡跑，回娘家吃午飯，向媽訴苦，談天，有時自己開車來，有時由安地開車送來。

日子一天一天過去，一年之後，安地已經開車送兩個人來串門子回娘家了。原來，他們不久就生了一個極其肥胖可愛的混血小女兒。

（04）

李太太見李教授對大小姐愛華的婚事不滿而怏怏不樂，心裡十分過意不去，眼見二小姐佩華，哥倫比亞大學也快畢業了，就特別留心，看看如何加點勁。

有一天，由菜場回家，正把車上大包小包的菜蔬往廚房內搬，發現李教授正在客廳看報紙，就對丈夫說："我今天在菜場買菜，你知道遇見了誰？"

"唔。"連頭都沒有到抬。

"遇見了另外一位李太太！"李太太很興奮的說。

"現在李是中國的第一大姓，也就是說，全世界以姓李的人數最多，遇見了姓李的有什麼稀奇。"她丈夫沒好氣地回答，繼續看他的報紙。

"我遇見的是那位好客的李太太呀，你知道他們為什麼那麼好客嗎？"李太太像發現新大陸一樣的高興。

"當然是錢多花不盡，飯多吃不完啦。"李教授打算不再理他太太了，昨天球賽結果雖然已經在電視上看過了，但是細節及內容，還是報紙報導得比較詳盡。

"你知道不知道！"李太太菜也不搬了。

"誰幫忙搬菜呀？"李大朝樓上大喊，聲音剛落地，樓上咚、咚、咚、下來三位如花似玉的姑娘們，忙著把菜搬到廚房裡去，她自己反倒一屁股坐在她丈夫旁邊的沙發上。

"那個好客的李太太呀，她說她家女兒多，在美國找中國女婿很不容易，一定要要父母留意、幫忙才行，所以經常在家中請客，請的都是些年輕有為的、未婚的中國男子，給自己的女兒們製造機會。"李太太壓低了嗓子，很神秘地說。

李教授一聽，果然動容，巴掌朝大腿上一拍，恍然大悟道："難怪他們每次請客，都沒有請我們，我還以為他們暴發戶不識斯文，不敢請我們這些有學問的人，原來是怕我們家漂亮的女兒們，把他們家的比下去了呢！"

李教授這一下子高興了，對這件事也就留心了起來。

二小姐佩畢平常在學校，住在學生宿舍內，忙著做實驗、寫論文、申請學校研究院，回來的機會極少。這次週末，李教授好不容易專程到哥大女生宿舍找她，把她帶回家來。老二回來后並不休息，整天在樓上自己房內做功課。

李教授費盡周章，把他自己教書的哥倫比亞大學全部中國研究生的名單及全部可以搜集的到的資料，包括照片，比做試驗還要專心費力，花了九牛二虎之力收集了起來，老倆口正在樓下廚房的飯桌上，戴起老花眼鏡，一個一個仔細研究呢！

只聽見李太太說：「這個長得體面，請了罷！」

「這個的父親是虎豹協會的會長，非請這個不可。」李教授說。

「這個看起來身體不錯，也可以請。」

「他的成績好，每年都在榮譽學生的名單上。」

佩華在樓上讀書讀倦了，想下來散散步，走動走動，就由房中出來了。她近來功課忙，頭髮長長的，沒空修剪，烏溜溜的一直長到腰下，配著她那吹彈得破的鵝蛋臉。眼鏡後面，有一雙長長的丹鳳眼，黑黑的眉毛，菱角似的紅唇，穿了件家常的寬大襯衫，緊身的牛仔褲，裹著修長的腿，高挑的身材，比時裝模特兒還要漂亮幾分。他看見父母正聚精會神，很熱心地討論著，連她進廚房都沒有覺察，便好奇的停了下來。

只聽見李太太說：「請客自助餐，你看要不要供酒？」

李教授說：「弄卓酒罷，他們都是研究生了，作個樣子，熱鬧些，隨他們愛不愛喝。」

佩華見父母要請客喝酒，忍不住插嘴道：「我的室友有個表哥，十分缺錢，也會調酒，可以找他來配酒，讓他賺點外快。」

(05)

李太太笑了起來，說：「傻孩子，我們中國人請客，那裡講究請專人調配，還不是擺卓酒，做個樣子罷了，那要特別請人呢？」

李教授說：「且慢，且慢，你說的那個人，會不會開車呀？」

「他有一輛老爺車，不過他十分能幹，老爺車一拋錨，他馬上就修好了。」佩華把這個窮光蛋形容得英雄似的。

李教授聽了，就對李太太說：「我們雇個幫手吧！住在紐約城內哥倫比亞大學大學的中國學生，都沒有車子，若他能接送客人，能調酒佈菜，再幫我搬卓重東西，倒也不失為一個幫手，要知我年紀到了，不比以前啦。」

佩華對父母的決定非常開心，給了父母大人一人一個香吻，飛快奔回樓上打電話去了。

她說：「好了，他只要六十元一天的工資。現在好了，他可以掙點錢，免於饑餓了。"

顯而易見的，我們這位李家二小姐佩華，早已把散步走動的事，拋到九霄雲外爪哇國去了。

中外習慣不同，李教授就不記得生平什麼時候曾經擁抱、吻過自己的雙親大人。不過，李教授夫婦對於佩華小姐的香吻，是十二分滿意的，兩人互相微笑，對看了一眼，又繼續討論客人名單。

李太千叮萬囑，要佩華週末一定要回家待客，那知佩華不但出乎意外地一口答應，而且還提早星期五半夜就很熱心地把室友的表哥一同帶來了。

這個幫忙的是個碧眼黃髮的美國白人，長得雖然還不錯，只是衣衫破爛，牛仔褲洗得發白，都起了毛邊，反正他只不過是請來的便宜的臨時工而已。李教授雖然皺了一下眉頭，也就不再多說什麼，立刻指揮他到地下室去搬重東西。

老美牛油麵包吃得足，搬重東西不費吹灰之力。這個工人，身手還算俐落。一教授年紀大了，有了這個幫手，省了不少力氣，覺得六十元花得實在值得。

下午三點鐘，李教授將客人的位址姓名，交給了這個洋青年。

"爸爸，它叫比爾。"二小姐佩華說。

比爾接過名單，到車內十分幹練的找到他要到的那張紐約市詳圖。

二小姐佩華陪他鑽進車裡，兩人一同研討地圖。

不一會兒，車子發出一聲轟然巨響，接著車尾冒出濃煙。原來，車子發動了，過了好一陣子，才噗噗地往前移動。

"停車，停車！"李教授一看不妙，忙跟在車后追著喊。

"爸爸，怎麼了？"佩華由車內伸出頭來，她把烏溜溜的長髮梳成了一個馬尾，扎在腦後，她的頭一動，馬尾就在腦後晃蕩，充滿了青春活力。

"汽車丟人現眼事小，把客人半路拋錨，豈不是誤了你的終身大事。"李教授說的是中文。

"什麼大事？"佩華用手推鼻樑上的眼鏡，滿臉問號。他在哥倫比亞大學修了二年中文，每年都得到 A，可惜書本上的中文與日常應用中的中文略有出入，但真正對話的時候，往往不太靈光。

"叫他把我的汽車開去罷！"李教授把自己車鑰匙給了比

爾。

(06)

　　"佩华，妳不必亲自去接客人，让他一人去就行了，你快去把自己打扮打扮吧。"看见佩华又打算跟着車子去了，李教授忙对女儿说道。

　　"到今晚不要戴这种學究式的眼镜了，反正又不看书。"李教授見二女儿很聽话地下车，忙又对佩华加了一句。

　　車子由長島到纽约来回两趟，才把客人接齐。

　　佩华平常忙于读书，不甚注重外表，天生美人胚子，今天穿了件苹果绿的洋装，同一颜色的平跟鞋，又用同色的絲带，在马尾上打了一个绿蝴蝶结，白淨的脸蛋儿，不施脂粉，要多美就有多美。

　　二小姐佩华眨著她的黑色，美丽，东方式的丹凤眼，没有戴眼镜！

　　"怎么样？"妈妈关心的问。

　　"不戴眼镜，每个人看起来都格外和善可亲。"李二小姐笑眯眯地回答。其实他近视并不怎麼深，不戴眼镜也不会出什么大差错。

　　当李太太与二小姐对话时，力教授看见他家四小姐黛华也穿戴整齐，全套功夫装备，也没有戴眼镜。

　　"黛华，你在这里做什么？"李教授很诧异地问

　　"爸爸，我才由邓叮的录影带中学会了一套全新的中国功夫，我要看这批中国客人中，有无能人可以较量较量。"李四小姐黛华道。

　　"你为什么不戴眼镜呢？"做爸爸问道。

　　"爸爸不是叫我们不要戴眼镜吗？我这不是十分听话？"黛华也眨眼睛。

　　"黛毕，这些饭菜有什么问题吗？你为什么每一道菜都要一个又一个的去闻呢？"开飯之后，李教授关心地问。

　　"黛华不是闻菜，她不过是一位近视的小姐，没有戴眼镜罢了。"做妈妈的回答道，知女莫如母，一点都没有错。

　　"哦？哦！"老爸突然大为紧张，但事已至此，为父的忙于待客，无暇细细解说为什么李四小姐需要戴眼镜。看见一屋子的精英，想到女儿前途，李教授打心里高兴起来。

　　"佩华，那个洋工人应该会照顾自己的，你大可以放心。你来，你来，我给你介绍王国栋，他的父亲是药学系的教授，德高望重。"佩华急忙像风摆杨柳一般的走过去王国栋握手。

　　"佩华，这是方砥柱同学，北京中国人民大学毕业的高材生。

你去倒杯酒给他。"佩华又到比尔那边，拿了杯酒过来。

"黛安，有什么不对吗？你为什么要逐一地嗅客人的脸呢？"李教授大惊失色。

"哦！她没有戴眼镜。"李教授自己回答了自己的问题。事已至此，夫復何言。

"佩华，比尔是我们花钱请来帮忙的，你不是早就从厨房端了汤给他喝过了吗？不必再一趟一趟尽拿东西给他吃了。"李教授说。

"来，来，来认识这位出生南京世家的张家驹同学。"老爸边说边拉了二小姐佩华过来。

(07)

打宴會一開始，李教授就緊張得不得了，一共四隻眼睛，只只派上用場。一隻盯牢四小姐黛華，生怕出了什麼差錯，兩隻眼睛要掃描全場，務使全部賓客注意到他家美麗大方的佩華，剩下的一隻，還要竭力使我們這位以二小姐明瞭，李家對於比爾的胃，是不負任何責任的。所以沒多久，我們這位中國老人家，就覺得疲憊不堪。

"黛華在哪裡？她又要出什麼點子呢？"過了一會兒，不見四小姐黛華的芳蹤，這可把我們的教授嚇壞了。

"她有點失望，這些中國留學生，都沒有聽過鄧叮的大名。"李太太告訴李教授。

現在，我們的中國老紳士，終於緩過一口氣來，因為四小姐黛華放棄了聞嗅賓客，逕自回到樓上香閨，對鄧叮的中國功夫，自行做更深一層的研究去了。

他老人家心中一塊大石落地，加以看見佩華心情很好，綠蝴蝶般地在一批中國未來的精英中周旋，做爸爸的真是笑得合不攏嘴來，不由得多喝幾杯。

有什麼關係，汽車的鑰匙在比爾那裡，沒有問題，比爾自會開車送他們回紐約。

未等客人全部散完，李教授已經倒在床上，呼呼大睡。

第二天早上醒來，床板空空如也，他忙披了睡袍，到廚房去，看見三個小女人正在用冷牛奶泡著早餐吃。

"媽媽在哪裡？"他忙問道。

"媽媽在地下室收拾，我們吃完早餐，也去幫忙。"老么范華用童稚的聲音的聲音嬌聲嬌氣地說道。

"爸爸，我們都在吃玉米片，你要不要也吃些？"老么范華問老爸。玉米片是玉米片壓幹了炸的，十分酥脆。

「不要，范華，爸爸不要吃。」他慌忙回答。

「不要，范華，絕對不要，這種美國式的早餐，咬在牙齒上蟋蟋作響，我的頭有點疼，受不了那種蟋蟋响聲。」他生怕老么真的倒玉米片給他吃，嚇壞了，不但連連搖頭，還將雙手亂搖，以加重語氣。

咖啡也等不及喝了，他走到地下室，看見太太正在收拾碗盤，也趕快過去幫忙。

「佩畢哪裡去了？」他問太太。

「佩畢功課忙，要趕回學校去，剛好那幫忙的工人拿到工錢，也回紐約，佩華搭他的便車走了。」李太太答。

「他們什麼時候走的？」李教授追問。

「大約清晨三奌左右罷！我也睏了，沒有看表。」李太太想了一下道。

難怪昨夜睡夢中彷彿夢見佩畢穿了中國新娘妝，鳳冠霞披，與這些少年中最好的一個，正在舉行中國式的婚禮，夢見中國式的鞭炮，撲撲作響，敢情是那臨時工，開走他的老老爺車，發出來的撲撲之聲。

「你看了佩華看中了哪一個？」李教授有點恨然，一面幫太太將盤子堆在一起，一面問正在擦桌子的李太太。

「她昨天對每個人都很客氣，至於對那個特別好，倒看不出來。」李太太仔細想了一下，如此回答。

李教授有點失望，沉默了半響。

「我們是不是白費力氣、白花錢了？」最後，他勉強問道。

「他們年輕人都在紐約，這一次大家認識了，以後見面大家就是朋友，做了朋友，就好發展了，認識是最重要的，他們以後一定會常常見面。」李太太怕丈夫失望，忙安慰丈夫道。

李教授覺得太太的話十分有理，第一步認識最重要，是扎基的工作，要發展，就要靠緣份了。

(08)

從此，李教授上班，除了教學之外，就是努力注意打聽佩華的動向，回家好與太太討論。

可惜打聽的結果，很讓人失望。那天請的客人中，不少人都對李二小姐有興趣，因為，娶了她也就是娶了美國公民，身份就有了著落，這是留學生最關心的事。但李二小姐佩華都推說功課忙，電影跳舞一概不幹。在學校餐廳里吃三明治倒是相遇了很多次，每次都是跟不同的人，很可能是吃飯的時候臨時遇見的。總而言之，畢業在即，不知道她一天到晚在哪裡用工，一點也沒有與中國男留學生談情說愛的跡象。

李教授夫婦也無可奈何。時代不同了，現在女孩終身大事，一不靠做媒，二不靠相親，家中還有五個未婚女兒，真急得父母親大人如熱鍋上的螞蟻。

有一天，要李教授回家，出乎意外，佩華比他先到家等他了，聽見他汽車回

來的聲音，如飛一般由屋內跑到車道上迎接他。

"好消息，好消息，爸爸，告訴你一個好消息。"

"你要結婚啦？這麼高興？"他問。

"好消息！爸爸，加州醫學院快要收我做學生了，要我去口試。"佩華手中

揚著一封信。

"那好極啦！口試就是談談罷了。"李教授自己在哥倫比亞大學教書，這些程序當然清楚得很，他們家的二小姐聰明伶俐、口齒清晰、口試教授哪會有什麼異議呢？

"什麼時候去呢？"李教授覺得十分依依不捨，這一去如同放鳥出籠，回來的機會，比她在紐約讀書時更少了。

"明天。"佩華回答道。她長長的眼睛發出興奮的亮光，我們的李二小姐是愈來愈漂亮了。

"那早點收拾罷！"李教授十分傷感，突然想到李太太，做母親的一定會更捨不得女兒長大離家，他急急要去安慰他的老妻，他們攜手過了半輩子。

第二天，天濛濛亮，他起床由視窗向外看，看見佩華揹了一個大背包，站在家門口等車，腳旁有一個更大的行李袋。她的母親李太太、四妹黛華也靜靜地站在那裡陪她。

李教授忙起身穿了睡袍，下樓送女兒。

"這麼早？"他說。

"爸，我們要早點出發，因為他的車子很舊了，怕萬一拋錨修理，耽誤時間，從紐約到加州要五天的路程，我怕誤了口試時間。這次學校放兩周假，時間比較充分，它也要我去熟悉熟悉環境。"

正說間，一陣撲撲的響聲，打斷了佩華的話。李教授覺得這撲撲之聲聽在耳裡，很是熟悉。一會兒，一輛破舊的老爺車在他

們門口停了下來。李教授看那車子，覺得似曾相識，車門一開，車內跳出一個漂亮的年輕人，又健康、又活潑，虎虎生風的樣子，也好像哪裡看見過，是了，是了，你看那洗得發白的牛仔褲，那邊都磨得發毛的牛仔背心，不正是那個金髮碧眼的臨時工比爾嗎？

李教授以為他在夢中，就咬了一下嘴唇，嘴唇發疼，不是做夢。又以為是眼睛發花，看錯了人，用力地把眼睛眨了幾眨，不錯，就是他。

"他來做什麼？"李教授詫異地問道。

"爸爸，他是比爾呀，爸爸，你忘了嗎？你還付過六十元給他呢！他今年哥倫比亞大學醫科畢業，要到加州醫學院去做實習醫生，我們至少有兩個禮拜可以在一起了。"

李教授只覺得眼前金星直冒，要昏倒。佩畢忙過來扶他，黛華奔進屋內端來一杯水交給李太太，要給他喝，比爾也急急忙忙由車內取出一個醫師用的黑色醫藥手袋來，給李教授診脈。

早晨第一線陽光，從東方樹梢中露了出來，李教授睜開眼睛，只見比爾那對透明的藍眼睛，沈著地看著自己，自己的手腕，被比爾強有力的大白手抓著，正在量脈膊呢！比爾那捲起袖子的手臂，反射著清晨的陽光，每一根金黃色的汗毛，都看得清清楚楚。

李教授很不情願地把頭掉過去，正遇見太太那對焦急的眼光。什麼時候她的眼角，也起了魚尾，她的頭髮也有幾許花白了呢？

(09)

果然不出李教授所料，好幾家著名的醫學院都願意收佩華做學生，也正如李教授的預測，佩華決定到加州三藩市去。

李教授夫婦帶了極矛盾的心情，帶了全家另外五個女兒以及一個混血的小外孫女兒，開了兩部汽車好，浩浩蕩蕩，參加佩華的畢業典禮。幸好比爾那天值班，所以整個典禮中除了李太太流了幾滴眼淚，李教授嘆了幾口氣之外，一切尚稱圓滿。

第二天，佩華與比爾，雙雙悄悄到市政府去領結婚執照，全家都鐵桶似的瞞著老爸，所以倒也相安無事。

佩華是李教授最寵愛的女兒，小時常常帶他去釣魚打球，也曾夢想她是個兒子，尤其希望他長大後能在自己的試驗室內做研究，繼承衣鉢。這些如今都化為泡影。現代男女平等，女兒結婚離家，雖然不是潑出去的水，但要收回來也是談何容易。兩夫妻同樣的悵然無奈，好在大小姐愛華常常帶了小寶寶回娘家串門子。

週日上班，長途往返，倍覺辛苦，所以週末，李教授都盡量睡到日上三竿。這天是星期六，刺目的陽光，鑽進嚴嚴關著的窗簾縫內，固執地射著他的眼睛，打擾了他的清夢。隔壁房內，震耳欲聾的音樂，強行敲打著他的耳鼓，她只得慢騰騰地從床上爬起來。

探頭到隔壁一看，五小姐依華與一位長著長長卷金髮叫做伊莎的女孩，把電視和音響，一起開得極響，兩人並排趴在地毯上，正出神欣賞呢！

"嗨！嗨！嗨！"四小姐房內傳出短促的呼聲。原來四小姐黛華已經決定要身體力行，不但要詳讀中國武俠小說、看中國武俠錄影帶，還要真槍真刀，努練習鄧叮的中國功夫武藝。

"最高的境界是刀槍不入，嗨！嗨！嗨！"四小姐黛華喊道。

李教授雙手扶著樓梯，直著脖子，小心下得樓來，生怕他的腦子會裂開。

進了廚房一看，么小姐范華正在一面吃東西，一面聚精會神地看著一本洋文童話書。李太太與大女兒正在說什麼，愛華臉上一分氣憤的顏色，顯然母親李太太正在安慰她。

"又是批評他那洋婆婆了。"他無聊地想。由太太那兒曉得大女兒和洋婆婆不和。他也懶得去管這些婆婆媽媽的芝麻綠豆小事，不過有時也隱隱約約地得意：

"你看，我說過洋人嫁不得罷！"

(10)

人人都沒有發現他下樓，當他那一人自己玩厭了的混血小外孫女，見李教授進得廚房，如獲至寶，用不清楚的嬰兒聲音喊道："公公，抱抱。"

她親了又親外孫女兒光光滑滑可愛的小臉蛋，又嗅著她身上甜甜的奶香味，小寶貝被外公又親又嗅，高興得咯咯直笑。

"可惜這麼可愛的小傢伙是混血。"李教授恨然地想。

正在此時，屋子外面，由遠而近，響起托的摩托車聲，由小而大。

李教授從廚房向窗外一看，一輛風馳電掣的摩托車突然剎住，停在他家門口的汽車道上，車上先跳下來一個身穿黑色皮夾克、黑皮褲子的年輕人，這人頭髮的顏色，有紅又有綠，根根都豎在頭上，顯然是用膠噴過。雖然分不清他頭髮原來的顏色，是個白人是無疑的。最奇怪的是他只帶了一隻長耳環，那耳環被他這么一跳，在耳上直搖晃。

這人自己跳下車后，又把車後一個穿著同樣黑皮套裝的女孩子，雙手由車上抱下，倆人難捨難分的親起嘴來。那女孩一隻耳上戴了耳塞，另一隻耳上戴了與那少年同一式的耳環，現在一人分帶一隻，兩隻耳環反射著週末五月長島的陽光，正在一同晃蕩。

李教授十分異，這對年輕人為什麼在他家門口親嘴，就把脖子特別伸長了些，向窗外仔細再看，只見二人四隻黑皮短指手套內，由十支是黃色的手指，另外十支卻是白色手指，露在黑皮手套外面，特別醒目。

他這裡正仔細端詳呢！他家老么范華突然發出一聲尖叫，右手的湯匙，左手的書，哐噹兩聲，全被她丟到地上的手上，李教授只覺得眼前晃過三個人影，范華已由他身邊衝過，向屋外飛奔，口內還在大叫：

"卡爾，卡爾，龍捲風卡爾！"

這時他太太與他的大女兒愛華早就興奮地跑到門外去了。

再看原來在樓上的五小姐依畢與金髮好友伊莎以及四小姐黛華，不知何時，都已經早就圍在那洋少年身邊，每人都拿了紙筆，有的還捧那少年的照片，個個口中怪叫，要這個叫卡爾的洋少年簽名呢！

那兩個親嘴的人，被這被問及一打攪，只得分開，那洋少年匆匆地給黛華、依華、伊莎以及范華她們簽完名。

不知何時，所有洋鄰居的門都打開了，每家的鶯鶯燕燕，都發了狂一般，奔到李家門口，要那洋少年簽名，還拼命扯他的衣裳，敲打他的車子，那少年似乎害怕極了這種場面，連忙發動了摩托車，這群雌之中殺出一條生路。

摩托車由近而遠，一溜煙逃得無影無蹤。正是：急急如喪家之犬，慌慌如漏網之魚。

只剩下一群發了狂的年輕女孩們在李家門口尖叫，過了好一陣子，才停了下來。洋女孩們也三三兩兩慢慢散去，一路上還興奮不已的吱吱喳喳，談個不休。

再看那位正在一家大門口出神的東方女孩，原來是李家三小姐彩華，她手中抱了吉他，正在發呆，看見爸爸李教授眼露凶光，一臉怒容，嚇得一溜煙跑到樓上，把房門反鎖，躺在床上，重溫昨夜與卡爾共度的舊夢去了。

(11)

李太太看見李教授頹然坐的沙發上，就由廚房端出了一杯熱茶給丈夫。李教授默然接了過來，口中慢慢地吹著杯中的茶葉。

"彩華學理科雖然不如她兩個姐姐，可是音樂方面是有些天才的，進了紐約大學的音樂系，現在在美國，可是行行出狀元的。"李太太輕輕地說。

李教授不響。

"上次請客的中國學生名單呢？還在不在？"沉默的好一陣子，李太太輕輕地問。

李教授點點頭。

"現在暑假開始了，他不能再住宿舍裡面，我們一定要好好地管教她，不許他胡來。"李太太小心翼翼地說。

"我們再請一次中國學生罷，這次絕對不許由外國人來調酒了。"又過了好一陣子，李太太又提議道。

三小姐彩華聽說這個暑假不許她外宿，氣得她在家哭哭啼啼，茶飯不思，鬧了好一陣子。這次請客，強迫她參加，被她一口拒絕。

"彩華，你聽我說，我們這次請客，專門為了妳，大家專來看你表演，唱中國歌，跳中國舞，妳說可好？"李太太告訴老三。

這一下，三小姐彩華果然破涕為笑。

"那我那我得好好練練。"彩華忙道。

"唱中國歌要有中國氣氛，我要用吉他彈出中國情調來。跳中國舞蹈，要好好查書，研究中國歌舞的基本精神。"李三小姐又說。

現在李三小姐彩華忙著查書，找道具，又去弄了音樂帶配音，還央求大姐和媽媽趕工做了一件小鳳仙裝，穿了新一在穿衣鏡前左左右右仔細端詳，比誰都熱心。

李教授夫婦這才把一顆懸得高高的心，放了下來。

"什麼時候才開派對呀？彩華天天練唱，把我們的耳膜都要震破了！"姐妹個個都在抱怨。

這次請客，李教授拼了老命，親自開車，滿載了一車由中國來留學的研究生，另一批由於一個叫趙萬里的中國留學生，向朋友的管家借了一部全新的卡迪萊客車，又大又穩，也滿滿地裝了一車，浩浩蕩蕩地由紐約市開到要長島李家門。

李家前們大開，李教授與李太太，笑容可掬，站在門口迎接賓客。

奇怪不奇怪，每位客人下車，都朝著李家客廳看著，個個表情古怪。主人李教授及李太太也就隨著眾人的眼光，朝自家廳中看去。

不看猶可，一看之下，真是啼笑皆非。

原來李家五小姐依華及她的金髮碧眼的好朋友伊莎，兩人各穿了一件黃色線衫，線衫上各有兩個大紅的中國字。李五小姐依華東方女孩的胸前的字是：「美味」。

白小姐伊莎那發育的飽滿的胸前的紅字，寫的是：「滋養」。

你看，李教授及李太太多難為情罷！

「趙萬里同學，請你代我們招呼一下客人，我們有點急事要處理，處理好馬上就回來。」李教授與李太大，忙把五小姐依華及女朋友伊莎趕到廚房。

在廚房裡，李教授要五妹把四姐找來，因為四小姐黛華識得一些中國字。

「教四姐黛華把這四個中國字的意思翻譯給你們聽，趕快把線衫脫了，從今以後，不許再穿這兩件衣服。」為父下了命令。

「什麼時候才開派對呀？彩華天天練唱，把我們的耳膜都要震破了！」姐妹個個都在抱怨。

「爸爸，我就在這裡。」黛華穿了另一套新的功夫裝，手握寶劍，笑容滿面。

「三姐彩華並不是唯一努力排練的人，我也一直在鍛煉的鄧叮大俠的獨門身手哩！」四小姐黛華對他那嚇壞了的老父母宣佈。

「這次我們請客，是為了三姐彩華，不是為了你呀！」可憐的李太太說道。

「哦！你們要為我們六姐妹每人請一次客呀！」黛華恍然大悟。

「我們不要中國來的研究生做我們的客人，我們要請與我們同樣年齡的客人。」五小姐依華嚷道，她今年還不到十五歲。

「對極了，我要請一批鄧叮影迷俱樂部的會員才是。」四小姐黛華振振有詞的說道。

李教授與李太太急於招呼等在客廳的客人，無暇與老四老五討價還價。從兩位小姐臉上滿意的表情的來推測，老爸老媽已經答應了她們的請求。而且，看來這些承諾，一定令二老有長期的後悔。第一，長島那有十五歲左右的中國男孩呢？再說，倒底什傢達是鄧叮影迷俱樂部？

事有輕重大小，做父母的得先應付眼前的燃眉之急，也就唯唯諾諾急忙趕到客廳。

這次宴會，幸好趙萬里性情活潑，前前後後端菜倒茶，幫了主人不少忙，大家熱熱鬧鬧，都吃了個飽。

(12)

　　高潮在飯後，彩華彈著吉他，有時獨唱，有時獨舞，有時全體合唱，有時大家用手敲拍子，真是賓主盡歡。

　　三小姐彩華穿了鳳仙裝，還特地仿民國初年的電影，梳了小鳳仙的髮結，不但歌聲甜美，舞姿曼妙，最令人敬佩的是她的組織能力，把在場的人，都引到這個家庭的狂歡中來了，自始至終，都沒有冷場。那個趙萬里又特別湊趣，手中的錄影機不停地咔咔作響，鏡頭快門一直對準了李家這位才藝雙全的三姑娘。

　　一曲既終，掌聲如雷，歡聲四起。

　　“李彩華，你真了不起，真是才藝雙全！”趙萬里特地倒了一杯可口可樂給李三小姐，還很誠懇地說。

　　彩華抬頭一見是他，她好像是他鄉遇故知，又像久別重逢，對他燦然地笑著，像一朵盛開的花朵一般。

　　“噢，趙萬里，你怎麼也來了呢？”彩華對他很親熱地問。

　　“對了，你是哥倫比亞大學歷史系的研究生，我家爸爸媽媽最喜歡中國來的男研究生的。”李三小姐彩華側著頭沉思了一會兒，恍然大悟。

　　“彩華，怎麼好久沒看見你了呢？”趙萬里也很親熱地問李家三姑娘。

　　“趙萬里，我家爸爸媽媽不許我暑假裡到紐約過夜了，他們像守犯人一樣的守著我，幸好可以練練中國歌舞，不然我就要發瘋了。”三小姐向他訴苦。

　　原來李彩華不但認識這個趙萬里，聽兩人對話的口氣，還很熟悉呢！李教授不由得不對他注意起來。

　　只見他高高瘦瘦，剪了個西裝頭，乾淨俐落，戴了一副黑邊眼鏡，一件深灰色的長褲，上面罩了一件新的藏青色的短袄，真如玉樹臨風，數不盡的風流瀟灑。中式的短襖，原來是可以這麼好看，眼鏡後面的眼睛，又是這麼多情，東方的男子，多麼迷人！

　　“爸爸，他叫趙萬里，哥倫比亞大學歷史研究所的高材生，專攻傳記文學，已經出版了一本書叫'天才的馬朋朋'，分上、中、下三集，集都出了好幾版，後來又發表了一本'牛友友傳'，剛印出來就轟動了，書被搶購一空。現在他正著手替目前美國最紅、最熱的歌迷偶像用最新的手法來寫一部新式的偶像傳記。”彩華見父親的眼光老是去注意趙萬里，就笑盈盈地對李教授說。

　　李教授正在上上下下打量這個年輕人，不意被女兒發現，本來有點訕訕地，現在見三姑娘如此大方，而且還這樣熱心介紹，真是喜出望外，笑逐顏開，忙叫太太拿蛋糕給趙萬里吃，倒茶給趙萬里喝。

曲終人散，已經深夜了。當晚趙萬里只能先送一部分學生回紐約城去，那另一部分留在李家過夜，次晨趙萬里回來，再送一次。

這次請客，圓滿結束。李家老夫婦也十分滿意，各各噓了一口大氣。因為心情好，所以雖然第二天發現彩華坐在趙萬里的身邊，跟著第二批學生也進了紐約城，兩人都沒有提出異議。

在「多謝老師！」，"再見"聲中，大家鞠躬揮手道別。

後來李教授特地打聽了一下，原來這趙萬里不但自己財富五鬥，家裡也是響噹噹的，爺爺是前清的翰林，老子官拜部長，舅舅是現在的省主席。

看來沒有什麼可以挑剔的啦！

李太太看老伴心情好，特地為丈夫燒了紅燒蹄膀、油亮的豆苗、辣椒香乾絲，都是李教授愛吃的。

大家喜孜孜地坐下。

"彩華呢？" 正要動筷子，李教授突然問。

"她進城去練唱了。打電話回來，說趕不及回來吃晚飯了，叫我們先吃。" 一太太也坐了下來。

李教授臉色沉了下來，大家默默吃了飯。

自此李教授特別注意彩華的行蹤。一連三夜都沒有回家。做老爸的忍無可忍，在請客單上找到了趙萬里的位址。

第四天一大早，按照地址到紐約去找。

找到了，在公園大道。

現在，李教授的日本平民車，停在那公園大道高級洋房前面。有一個好穿了筆挺制服的黑人，前來問他找誰。這制服是白上衣，紅長褲，長褲兩邊都有兩條筆直的白線。

"我找公園大道 250 號，第三零五公寓的趙萬里。" 李教授理直氣壯地說。

那穿制服的黑人，點了點頭，做了一個手勢，請他出來，他一腳踏出車子，那人就一屁股坐上車，關了車門，很熟練地把車子開走了。

"既來之，則安之。" 李教授想。自己又不是沒有見過世面的人，何況車子已被人開走了•

他在二零五號門前按了門鈴，另一個穿了同樣制服的洋白人，前來開門。

"我找二零五號 第三零五公寓的趙萬里。" 那白人也點點頭把他引進一個電梯，給他按了第三樓。

在進電梯之前，他注意到腳下的地毯又厚又結實，大廳中由高高的天花板上垂下的水晶燈，又大又閃閃發光，一切都十分豪華，十二萬分氣派。

電梯在三樓停下，他由梯內走出，到了 305 公寓再按門鈴，一個穿藍制服管家模樣男人過來開門。

"我找趙萬里，他住在這裡嗎？"李教授機械地重複著說了三次的話。

"趙萬里先生是我家主人的客人，正在為我家主人寫傳記，所以今年都要住在這裡。"那人說。

"那他在哪裡？我要見他。"李教授堅持道。

(13)

他在二零五號門前按了門鈴，另一個穿了同樣制服的洋白人，前來開門。

"我找二零五號，第三零五公寓的趙萬里。"那白人也點點頭，把他引進一個電梯，給他按了第三樓。

在進電梯之前，他注意到腳下的地毯又厚又結實，大廳中由高高的天花板上垂下的水晶燈，又大又閃閃發光，一切都十分豪華，十二萬分氣派。

電梯在三樓停下，他由梯內走出，到了 305 公寓再按門鈴，一個穿藍制服管家模樣男人過來開門。

"我找趙萬里，他住在這裡嗎？"李教授機械地重複著說了三次的話。

"趙萬里先生是我家主人的客人，正在為我家主人寫傳記，所以今年都要住在這裡。"那人說。

"那他在哪裡？我要見他。"李教授堅持道。

"對不起，密斯脫趙現在出去了，你可以把您的姓名、位址以及找他的事由有給我，我可以替你轉交。"那人很恭敬地說。

"他什麼時候回來？"李教授大聲地問。

"他平常很晚才回來，中午不回來吃飯，白天在圖書館工作。"那人很客氣地說。

"什麼圖書館？我要馬上見他。"李教授十分不耐煩，聲音更大更響。

"什麼事這樣大呼小叫的？"他們的的談話聲，驚動到裡面一個年輕的白人，顯然是主人，這人從臥室出來，穿了一件短褲，裸露著結識的上身，懷裡還摟著一個漂亮的東方佳麗。那美女上身著睡衣，裸露著美麗均勻的大腿，那不是彩華，又是誰？

再仔細一看，這個年輕白人不正是那不男不女的龍捲風卡爾嗎？頭茄現在變成銀色和藍色了。

李教授氣憤填膺，正想過去教訓教訓卡爾，又想要打彩華，巴掌才舉起來，才想到在美國父母是不許體罰孩子的，正在此時，不知何處出來一白一黑兩名大漢，上來把他的左右手各各抓住。

"爹地！"彩華一見，連忙掙脫卡爾的懷抱，來保護父親。

"你們退下去，他是彩華的父親。"卡爾見狀，一擺手，對那兩名大漢說。

"馬修，你去把中國茶具拿出來，衝點上等的中國茶招待李博士、李教授。他是有學問的人，你們不要怠慢了。"他對管教說，原來這管家的名字叫馬修。

"我要去告你！"李教授喃喃地說。

"彩華已經二十一歲，早就過了合法年齡，我今年才二十歲，看誰去告誰罷！"卡爾胸有成竹地轉向李教授說道。

李教授當然不肯飲茶，儘管它實在口渴。他看也不看彩華一眼，由樓上衝進電梯。

(14)

後來，他怎麼回到車上，又如何開車回到長島家中，什麼都記不得了。

進了客廳，裡面光線不好，黑黑暗暗地，他走過去拉窗簾的時候，突然發現黑暗裡有人坐在沙發上。

此人是誰呢？為何坐在黑暗中呢？

"爸爸！我要與艾安地離婚！"愛華早就坐在客廳裡專門等待老爸了，見老爸過來拉窗簾，不由得傷心地大哭起來。

"慢來，慢慢來！我們中國人講究嫁雞隨雞，嫁狗隨狗，艾安地犯了什麼大法，非要離婚？不要哭，千萬不要哭泣，我們從從長計議，爸爸會給你做主！"李教授慌了手腳。

只有李太太能應付這種場面，可是，她哪兒去了呢？

"爸爸，他不愛我了，他愛他媽媽！"愛花哭得如梨花帶雨，氣都喘不過來。

"怎麼回事呢？"

"我洋婆婆，天天到我們家來看小孫女，一天到晚又親又抱，老年人有細菌呀！"罪狀之一。

"她說她希望再有一個孫子，這不是重男輕女嗎？"罪狀之二。

「她說小寶寶什麼都好看，若是鼻子再挺一臾就好了，這不是種族歧視嗎？」罪狀之三。

「她天天來我家，我們一臾隱私權都沒有了。」罪狀之四。

「我忍無可忍，勸她多休息，不必天天到我家來，她竟哭了起來，這不是不通情理嗎？」罪狀之五。

「她現在常常要安地到她那裡去，安地現在常常不在家了。」罪狀之六。

「今天又打電話來說，支票算來算去，收支不會平衡，要安地帶個小計算機去，替他媽媽算算。」罪狀之 N。

「小寶寶呢？」外公問。在美國，人人都說東方女子溫柔體貼，東方媳婦孝敬服從，他有什麼話說。

安地與我吵架，把小寶貝嚇哭了，有傷兒童心理健康，所以媽媽帶她出去吃霜淇淋了。罪狀之 N+1。

「叮，咚！」樓上傳出東西摔破的聲音。

「叮，咚！」四小姐戴華房內有什麼東西摔破了。

李教授與大女兒愛華的對話，至此暫停。響聲既然從四小姐黛華房內傳出，父女兩人就急急奔往樓上。

四姑娘黛華房內一片混亂，座燈倒在床上，椅子打翻，四腳朝天，所有的功夫錄影帶、CD、武俠小說，全部打翻在地板上。

(15)

四小姐跪在地上，眯著眼睛，雙手滿地亂摸。

「黛華，你丟了什麼東西嗎？」大小姐愛華問。

「你在找什麼呢？」爸爸也問。

「我們武術教練叫我們戴好隱形眼鏡再練功夫。可是沙袋打在臉上，把我的隱形眼鏡震得掉出了眼睛。」黛華很絕望地回答。

那懸在房間正中央的沙袋，仍然在不停地晃蕩。

「您們知道怎麼在這亂堆中尋找兩個小小透明的隱形眼鏡嗎？」李教授發愁地問道。

正在這不可開交的時候，樓下傳來汽車進車房的聲音。

李太太回來了！

老爸與大姐都急忙轉身，只有四小姐黛華還留在房中，不知如何是好。

「媽媽！」愛華首先奔下樓去迎接媽媽及女兒。

做爸爸的噓了一口氣，救星駕到，心中大石落地。李教授決定要回到自己房間，他急需獨自一人，冷靜地想想這一天所碰到的事。

在書房的書桌前坐下來，看見書桌上好大一堆信件沒有拆開，他嘆了一口氣，為了三小姐彩華，好幾天的信件都沒有處理。李教授隨手取過一封信，用裁紙刀裁開了信封，抽出信紙，看了起來。

突然，他腦子轟然一聲巨響，幾乎昏了過去。

原來信上說，法蘭西絲懷孕了，想要墮胎，但目前有技術、醫藥、健康上的困難，必須通知父母...。

奇怪，不是在美國墮胎，目前是不需通知父母的嗎？可惜自己平常太忙，沒有注意過這種問題，再說，李家么女范華結實健康...。

范華那可愛童稚的臉，在他眼前浮現，怎麼可能呢？范華才十二歲！

且慢，且慢，他把信仔細再重讀了一遍，怎麼信上女孩的名字叫法蘭西絲，他家么女的英文名字叫做法蘭呀！

再看，先是寫給一位姓 Lu 的先生，並不是給 Li 先生的，那不是盧先生嗎？那裡是李先生呢？

他急忙由椅子上跳起來，把頭埋到廢紙簍中，去找這封信的信封。

終於找到了！

地址是蘋果巷七號。

你看，洋郵差多糊塗！李家明明是白楊巷五號。Lu 與 Li 雖然只有一個字母相差，但姓李與姓盧是完全不一樣的呀！他們這社區，什麼時候搬來一家姓盧的中國人家呢？怎麼沒有聽見太太提起過呢？

這樣的中國人家，不提也罷！

他笑了，所謂怒極而笑也。

李太太及愛華在車房裡聽見李教授狂笑的聲音，心覺有異，愛華忙由車房奔到書房，李太太抱了小外孫女，也急忙跟了過來。他們打開書房一看，只見李教授因為三四天沒有的睡覺，滿眼紅絲，現在嘴又大大張開，高聲怪笑。

「公公，怕怕！」他的混血小外孫女，見到外公那樣子，嚇得哭了起來。樓上的女兒們，聽見爸爸在書房裡又哭又笑，也由樓上奔下來到書房裡看個究竟。

四小姐黛華戴了她酒杯底的眼鏡，大圈圈裡有小圈圈。

五千依華與她金髮碧眼的好友伊莎依命將那兩件黃線衫藏之高閣，再也不曾穿過。今天她們一人穿了一件綠線衫，每人衫前有兩個藍顏色的中國字，李依華那平坦小巧的胸前是："物美"。

伊莎那高聳發達的胸前，寫的是的：「價廉」。

"爸爸，我長大了要聽話，絕對不嫁白人丈夫，我長大了做一個女同性戀的人！"五小姐依華為了安慰老爸，獻殷勤地走到爸爸面前，對爸爸說。

（全文完）

中國天津〈老年文學海內外徵文獎〉

Family With Six Treasures

家有六仟金

作　者/余國英（Gwen Li）

www.amazon.com/author/gwen.li

出版者/美商 EHGBooks 微出版公司

發行者/美商漢世紀數位文化公司

臺灣學人出版網：http://www.TaiwanFellowship.org

地　　址/106 臺北市大安區敦化南路 2 段 1 號 4 樓

電　　話/02-2701-6088 轉 616-617

印　　刷/漢世紀古騰堡®數位出版 POD 雲端科技

出版日期/2021 年 3 月

總經銷/Amazon.com

臺灣銷售網/三民網路書店：http://www.sanmin.com.tw

三民書局復北店

地址/104 臺北市復興北路 386 號

電話/02-2500-6600

三民書局重南店

地址/100 臺北市重慶南路一段 61 號

電話/02-2361-7511

全省金石網路書店：http://www.kingstone.com.tw

定　　價/新臺幣 750 元（美金 25 元 / 人民幣 180 元）